Philosophy of Science

An Introduction

Thomas J. Hickey

ISBN 13: 978-0-578-57511-7 (hardcover)
ISBN 13: 978-0-578-60491-6 (paperback)

Converted and distributed by **www.ebookit.com**

Sixth Edition

Abstract

This concise and accessible book is a synthesis of the basic principles of the contemporary pragmatist philosophy of science. It discusses the aim of basic science, the methods of scientific discovery, the criteria for scientific criticism, and the nature of scientific explanation. Included is a description of a newly emergent specialty called computational philosophy of science, in which computerized discovery systems create and test new scientific theories.

It also examines the essentials of the underlying pragmatist philosophy of language that has made philosophy of science a coherent and analytical discipline, and that has given new meaning to such key concepts as theory, observation and explanation.

Preface

This book sets forth the elementary principles of the contemporary pragmatist (a.k.a. New Pragmatist or neopragmatist) philosophy of science including its underlying neopragmatist philosophy of language, and it briefly describes the new and emerging area of computational philosophy of science. Computational philosophy of science is not something outside of philosophy of science; it is twenty-first century philosophy of science. Obviously this brief introduction can make no claim to completeness.

My previous books include ***Introduction to Metascience: An Information Science Approach to Methodology of Scientific Research*** (1976), ***History of Twentieth-Century Philosophy of Science*** (1995), and the e-book ***Twentieth-Century Philosophy of Science: A History*** (2016). This introductory book is also available as an e-book. The e-books are also available in a free Internet web site www.philsci.com, which offers the books as free pdf downloads.

In his magisterial ***Types of Economic Theory*** Wesley Clair Mitchell, Columbia University American Institutionalist economist and founder of the prestigious National Bureau of Economic Research wrote that the process that constitutes the development of the social sciences is an incessant interaction between logically arranged ideas and chronologically arranged events. Since modern science is an evolving cultural institution, this memorable Institutionalist refrain can be adapted for philosophy of science: The process that constitutes the development of philosophy of basic science is an episodic interaction between analyses in philosophy and developments in science. Modern philosophy was formed in response to the historic Scientific Revolution commencing with Copernicus and completed by Newton. **In fact all positivist philosophy of science could be called a footnote to Newton.**

Since the demise of positivism, there has been an institutional change in philosophy of science. The institution-changing developmental episodes that produced the contemporary pragmatist philosophy of science are the scientific revolutions created by Einstein and especially by Heisenberg. These revolutions produced basic revisions in philosophy of language as well as in physics. **In fact contemporary pragmatist philosophy of science could be called a footnote to Heisenberg.**

Quine said that there are two kinds of philosophers: Those who write philosophy and those who write history of philosophy. Most books on philosophy of science treat philosophy of science in an historical perspective. This book is not a history of philosophy of science; it is a work in philosophy – a coherent and synthetic examination of the contemporary ascendant pragmatist philosophy of science. Both during and before the positivist era the purpose of philosophy of science was typically viewed in terms of justifying the superior epistemic status of empirical science.

Today few philosophers of science perceive any imperative for such justification of science, and often dismiss such efforts as merely pedantic exercises. Now the aim of philosophy of science is seen to characterize the practices that have made the empirical sciences so unquestionably successful. As stated below in the first chapter of this book: "The aim of contemporary pragmatist philosophy of science is to discover principles that explain successful practices of basic-science research, in order to advance contemporary science by application of the principles" (Section **1.01**).

This book has benefited greatly from my more than thirty years of practical research experience as an econometrician working in both business and government. This practical work has included application of my **METAMODEL** mechanized discovery system for most of my career. I have therefore concluded that the philosopher of science must get his hands dirty with data, in order to have a realistic appreciation of empirical science.

And I add that contrary to my early expectations I can now say in retrospect that the contemporary pragmatist philosophy has been more enabling for my successful practical economic research than my graduate-level textbook lessons in economics. I find economic theory has too much imputed fantasy and too little investigative empiricism. I hope this short introductory book benefits other working scientists as well as academic philosophers of science.

I expect that the reader may have the same difficulty assimilating this introductory book that I have had in writing it. The contemporary pragmatist philosophy set forth herein is an integrated system of inter-related concepts that are mutually defined by the context constituting the metatheory. Its exposition therefore cannot simply be linear, because any beginning presupposes concepts that follow. My attempt to cope with this circularity has been to approach the system in a sequence of levels of presentation. This treatment of circularity has occasioned some repetition in the exposition and some overlap among the chapters, in order to provide context for understanding. I have made the table of contents more detailed than the brief outline below.

Chapter 1 is definitional: it sets forth several strategic concepts used throughout the book.

Chapter 2 is didactic: it contrasts the basic features of neopragmatism with comparable ideas in the older romantic and positivist philosophies.

Chapter 3 is essential: it describes the new contemporary pragmatist (or neopragmatist) philosophy of language that is central to and distinctive of the neopragmatist philosophy of science.

Chapter 4 is synthesizing: it describes the four functional topics that are characteristic of basic-research science in terms of the new philosophy of language.

A reviewer of an earlier edition called this book a "manifesto". The book is explicitly addressed to academic philosophers and their students, and it does indeed advocate both the contemporary pragmatist philosophy of science and the new specialty called computational philosophy of science. Thus it has an explicit agenda. It furthermore proposes to benefit many academics in the philosophically retarded social sciences. Therefore this brief book *Philosophy of Science*: *An Introduction* might well be construed as a contemporary pragmatist philosophy-of-science manifesto.

TJH
October 2019

Contents

Chapter 1. Overview

Both successful science and contemporary philosophy of science are pragmatic. In science, as in life, realistic pragmatism is what works successfully. This introductory book is a concise synthesis of the elementary principles of the contemporary pragmatist (a.k.a. neopragmatist) philosophy of science, the philosophy that the twentieth century has bequeathed to the twenty-first century. This chapter defines some basic concepts.

1.01 Aim of Philosophy of Science

The aim of contemporary pragmatist philosophy of science is to discover principles that explain successful practices of basic-science research, in order to advance contemporary science by application of the principles.

The principles are set forth as a metatheory, which is sketched in this book. Basic science creates new language: new theories, new laws and new explanations. Applied science uses scientific explanations to change the real world, e.g., new technologies, new social policies and new medical therapies. Philosophy of science pertains to basic-science practices and language. However, applied science in the real world offers a positive feedback to basic science. Academics often have their pet ideas that motivate them to reject alternative but empirically warranted laws. But successful application of the scientific law will eventually influence even the most obdurate naysayer making him a pragmatic realist in spite of himself.

1.02 Computational Philosophy of Science

Computational philosophy of science is the design, development and application of computer systems that

proceduralize and mechanize productive basic-research practices in science.

Philosophers of science can no longer be content with more hackneyed recitations of the Popper-Kuhn debates of half a century ago, much less more debating ancient futile ethereal metaphysical issues such as realism vs. idealism.

Contemporary philosophy of science has taken the computational turn. Mechanized information processing (a.k.a. "artificial intelligence") has permeated almost every science, and is now intruding into philosophy of science. Today computerized discovery systems facilitate investigations in philosophy of science in a new specialty called "computational philosophy of science".

The pragmatist philosophers Charles Sanders Peirce (1839-1914) and Norwood Russell Hanson (1924-1967) had described a **nonprocedural** analysis for developing theories. Peirce called this nonprocedural practice "abduction"; Hanson called it "retroduction". Today in computational philosophy of science **procedural** strategies for developing new theories are coded into computer systems.

1978 Nobel-laureate economist Herbert Simon (1916-2001), a founder of artificial intelligence, called such systems "discovery systems". In the 1970's Hickey (1940) in his *Introduction to Metascience: An Information Science Approach to Methodology of Scientific Research* (1976) called the mechanized approach "metascience". In the 1980's philosopher of science Paul Thagard (1950) in his *Computational Philosophy of Science* (1988) called it "computational philosophy of science", a phrase that is more descriptive and therefore will probably prevail.

Mechanized simulation of successful developmental episodes in the history of science is typically used to test the plausibility of discovery-system designs. But the proof of the pudding is in the eating: application of computer systems at the frontier of a science, where prediction is also production in order to propose new empirically superior theories, further tests the systems. Now philosophers of science may be expected to practice what they preach by participating in basic-science research to produce empirically adequate contributions. Contemporary application of the discovery systems gives the philosopher of science a participatory and consequential rôle in basic-science research.

1.03 Two Perspectives on Language

Philosophy of language supplies an organizing analytical framework that integrates contemporary philosophy of science. In philosophy of language philosophers have long distinguished two perspectives called **"object language"** and **"metalanguage"**.

Object language is discourse about nonlinguistic reality including domains that the particular sciences investigate as well as about the realities and experiences of ordinary everyday life.

Metalanguage is language about language, either object language or metalanguage.

Much of the discourse in philosophy of science is in the metalinguistic perspective. Important metalinguistic terms include "theory", "law", "test design", "observation report" and "explanation", all of which are pragmatic classifications of the uses of language. For example in the contemporary pragmatist philosophy a "**theory**" is a universally quantified hypothesis *proposed* for empirical testing. A "**test design**" is a universally quantified discourse <u>*presumed*</u> for the empirical testing of a theory in order to identify the subject of the theory independently of the theory and to describe the procedures for performing the test. The computer instructions coded in discovery systems are also metalinguistic expressions, because these systems input, process and output object language for the sciences.

1.04 Dimensions of Language

Using the metalinguistic perspective, philosophers analyze language into what Rudolf Carnap (1891-1970) called "dimensions" of language. The dimensions of interest to neopragmatist philosophers are **syntax, semantics, ontology,** and **pragmatics.**

<u>Syntax</u> refers to the structure of language. Syntax is arrangements of symbols such as linguistic ink marks on paper, which display structure. Examples of syntactical symbols include terms such as words and mathematical variables and the sentences and mathematical equations constructed with the terms.

Syntactical rules regulate construction of grammatical expressions such as sentences and equations out of terms, which are usually arranged by concatenation into strings.

<u>Semantics</u> refers to the meanings associated with syntactical symbols. Syntax without semantics is literally meaningless. Associating meanings with the symbols makes the syntax "semantically interpreted".

Semantical rules describe the meanings associated with elementary syntactical symbols, *i.e.* terms. In the metalinguistic perspective belief in the truth of semantically interpreted universally quantified sentences such as the affirmation "Every crow is black" enables sentences to function as semantical rules displaying the complex meanings of the sentences' component descriptive terms. Belief in the statement "Every crow is black" makes the phrase "black crow" redundant, thus displaying the meaning of "black" as a *component part* of the meaning of "crow". The lexical entries in a unilingual dictionary are an inventory of semantical rules for a language. This is not "rocket science", but there are philosophers who prefer obscurantism and refuse to acknowledge componential semantics.

Ontology refers to the aspects of reality described by semantically interpreted sentences believed to be true, especially belief due to experience or to systematic empirical testing. **This is the thesis of ontological relativity.** Ontology is typically of greater interest to philosophers than to linguists.

Semantics is knowledge of reality, while ontology is **reality as known**, *i.e.* semantics is the perspectivist signification of reality, and ontology is the aspects of reality signified by semantics. Ontology is the aspect of mind-independent reality that is cognitively captured with a perspective revealed by semantics.

Not all discourses are equally realistic; the semantics and ontologies of discourses are as realistic as they are empirically adequate. Since all semantics is relativized and ultimately comes from sense stimuli, there is no semantically interpreted syntax of language that is utterly devoid of any associated ontology. If all past falsified explanations were simply unrealistic, then so too are all currently accepted explanations and all future ones, which are destined to be falsified in due course. Better to acknowledge in all explanations the degree of realism and truth that they have to offer. Scientists recognize that they investigate reality and are motivated to do so. Few would have taken up their basic-research careers had they thought they were merely constructing fictions with their theories or fabricating semantically vacuous instrumentalist discourses.

Pragmatics in philosophy of science refers to how scientists use language, namely **to create** and **to test theories**, and thereby **to develop scientific laws used in test designs and in scientific explanations.** The pragmatic dimension includes both the semantic and syntactical dimensions, such that the dimensions of language are telescoped.

1.05 Classification of Functional Topics

Basic-science research practices can be classified into four essential functions performed in basic research. They are:

1. Aim of Basic Science

The institutionalized aim of basic science is the culturally shared aim that regulates development of explanations, which are the final products of basic-scientific research. The institutionalized views and values of science have evolved considerably over the last several centuries, and will continue to evolve episodically in unforeseeable ways with future advancements of science.

2. Discovery

Discovery refers to the processes of constructing new theories. Pragmatists define theory language pragmatically, *i.e.*, functionally.

A theory is a universally quantified hypothesis that is proposed for empirical testing.

Today scientific discovery is facilitated by artificial-intelligence discovery systems.

3. Criticism

Criticism refers to the decision criteria for the evaluation of theories. Pragmatists accept ***only*** the empirical criterion. **The pragmatics of theory language is empirical testing** and it uses *modus tollens* conditional deductive argument form, which includes universally quantified statements and/or equations that can be schematized in nontruth-functional conditional form and that are proposed hypotheses. The scientific theory in the deduction is a set of one or several universally quantified related statements expressible jointly in conditional form.

Some linguists might regard the fundamental nontruth-functional conditional schema to be the "deep structure" of the language of basic science, while they regard the formulations of theory actually used by scientists as "surface structure".

A test design is a universally quantified discourse that is presumed for empirical testing a theory, in order independently to identify the subject of the theory and to describe the procedures for performing the test.

A **scientific law** is a former theory that has been tested with a nonfalsifying outcome.

4. Explanation

An explanation is language that describes the occurrence of individual events and conditions that are caused by the occurrence of other described individual events and conditions according to law statements.

Explanation in science uses *modus ponens* conditional deductive argument form, which includes universally quantified related statements and/or equations that can be schematized in nontruth-functional conditional form and that are scientific laws. A scientific law is a former theory that has been tested with a nonfalsifying outcome. Whenever possible the explanation is predictive of future events or of evidence of past events.

As with criticism, so too for explanation the nontruth-functional conditional schema may be regarded as the fundamental "deep structure" of the language of basic science, while the formulations of theory actually used by scientists may be regarded as "surface structure".

1.06 Classification of Modern Philosophies

Twentieth-century philosophies of science may be classified into three generic types. They are **romanticism, positivism** and **pragmatism.** Romanticism is a philosophy for social and cultural sciences. Positivism is a philosophy for all sciences and it originated in reflection on Newtonian physics. Contemporary pragmatism is a philosophy for all sciences, and it originated in reflection on quantum physics.

Each generic type has many representative authors advocating philosophies expressing similar concepts for such metalinguistic terms as "theory", "law" and "explanation". Philosophies within each generic classification have their differences, but they are much more similar to each other than to those in either of the two other types. The relation between the philosophies and the four functional topics can be cross-referenced as follows.

Chapter 2. Modern Philosophies

This second Chapter sketches three generic types of twentieth-century philosophy of science in terms of the four functional topics. Philosophy of language will be taken up in Chapter 3. Then all these elements will be integrated in a detailed discussion of the four functional topics in Chapter 4.

2.01 Romanticism

Romanticism has effectively no representation in the natural sciences today, but it is still widely represented in the social sciences including economics and sociology. It has its roots in the eighteenth-century German idealist philosophers including notably Immanuel Kant (1770-1831), progenitor of romanticism, and especially Georg Hegel (1724-1804) with the latter's emphasis on ideas in social culture. The idealist philosophies are of purely antiquarian interest to most philosophers of science today.

Romantics have historically defaulted to the positivist philosophy for the natural sciences, but they reject using the positivist philosophy for the social sciences. Romantics maintain that there is a fundamental difference between sciences of nature and sciences of culture.

Aim of science

For romantics the aim of the social sciences is an investigation of culture that yields an "interpretative understanding" of "human action", by which is meant explanation of social interactions in terms of intersubjective mental states, *i.e.*, shared ideas and motives, views and values including the economists' rationality postulates, that are culturally shared by members of social groups.

This concept of the aim of science and of explanation is a "foundational agenda", because it requires reduction of the social sciences to a social-psychology foundation, i.e., description of observed social behavior by reference to intersubjective social-psychological mental states.

Discovery

For romantics the creation of "theory" in social science may originate either:

(1) in the social scientist's introspective reflection on his own ideas and motivations originating in his actual or imaginary personal experiences, which ideas and motives are then imputed to the social members he is investigating, or

(2) in empirical survey research reporting social members' overtly expressed verbally intersubjective ideas and motivations.

Romantics say "social theory" is language describing intersubjective mental states, notably culturally shared ideas and motivations, which are deemed the causes of "human action". Some romantics call the imputed motives based in introspective reflection "substantive reasoning" or "interpretative understanding". But all romantic social scientists deny that social theory can be developed by data analysis exclusively or by observation of overt behavior alone. Romantics thus oppose their view of the aim of science to that of the positivists' such as the sociologist George Lundberg (1933) and the behavioristic psychologist B.F. Skinner (1904-1990). Romantics say that they explain consciously purposeful and motivated "human action", while behaviorists say they explain publicly observable "human behavior".

Criticism

For romantics the criterion for criticism is "convincing interpretative understanding" that "makes substantive sense" of conscious motivations, which are deemed to be the underlying "causal mechanisms" of observed "human action".

Causality is an ontological concept, and nearly all romantics impose their mentalistic ontology as the criterion for criticism, while making empirical or statistical analyses at most optional or supplementary.

Furthermore many romantic social scientists **require** as a criterion that a social theory must be recognizable in the particular investigator's own introspectively known intersubjective personal experience. In Max Weber's (1864-1920) terms this is called *verstehen*. It is the thesis that empathetic insight is a necessary and valuable tool in the study of human action, which is without counterpart in the natural sciences. It effectively makes all sociology what has been called "folk sociology".

Explanation

Romantics maintain that only "theory", *i.e.*, language describing intersubjective ideas and motives, can "explain" conscious purposeful human action.

Motives are the "mechanisms" referenced in "causal" explanations, which are also called "theoretical" explanations. Observed regularities are deemed incapable of "explaining", even if they enable correct predictions.

Some formerly romantic social scientists such as the institutionalist economist Wesley C. Mitchell (1874-1948) and the functionalist sociologist Robert K. Merton (1910-2003) have instead chosen to focus on objective outcomes rather than intersubjective motives. This focus enables the testability and thus the scientific status of sociology. But the focus on objective outcomes still represents a minority view in academic social science.

2.02 Positivism

Since the later twentieth century positivism has been relegated to the dustbin of history. Its origins are in the eighteenth-century British empiricist philosophers including John Locke (1632-1704) and most notably David Hume (1711-1776). But not until the late nineteenth century did positivism get its name from the French philosopher Auguste Comte (1798-1857), who also founded sociology.

The "neopositivists" were the last incarnation of positivism. They attempted to apply the symbolic logic fabricated by Bertrand Russell (1872-1970) and Alfred Whitehead (1861-1947) early in the twentieth century, because they had mistakenly fantasized that the Russellian truth-functional symbolic logic can serve philosophy, as mathematics has served physics. They are therefore also called "logical positivists".

Contrary to romantics, positivists believe that all sciences including the social sciences share the same philosophy of science. They therefore reject the romantics' dichotomy of sciences of nature and sciences of culture.

The positivists' ideas about all four of the functional topics in philosophy of science were influenced by their reflections upon Newtonian physics.

Aim of science

For positivists the aim of science is to produce explanations having objectivity grounded in "observation language", which by its nature describes observed phenomena.

Their concept of the aim of science is thus also called a "foundational agenda", although the required foundation is quite different from that of the romantics. But not all positivists were foundationalists. Otto Neurath's famous antifoundational boat metaphor once referenced by Quine compares us to sailors who must rebuild their ship on the open sea when we are unable to break it down in dry-dock. Neurath was a member of the Vienna Circle positivists.

Discovery

Positivists believed that empirical laws are inferentially discovered by inductive generalization based on repeated observations. They define empirical laws as universally quantified statements containing only "observation terms" describing observable entities or phenomena.

Early positivists such as Ernst Mach (1826-1916) recognized only empirical laws for valid scientific explanations. But after Einstein's achievements neopositivists such as Rudolf Carnap (1836-1970) recognized hypothetical theories for valid scientific explanations, if the theories could be linguistically related to language used to report the relevant observations. Unlike empirical laws, theories are not produced by induction from repeated singular observations.

Neopositivists believed that theories are discovered by creative imagination, but they left unexplained the creative process of developing theories. They define theories as universally quantified statements containing any "theoretical terms", *i.e.*, terms not describing observable entities or phenomena.

Criticism

Positivists' criterion for criticism is publicly accessible observation expressed in language containing only "observation terms", which are terms that describe only observable entities or phenomena.

The later positivists or neopositivists maintain that theories are indirectly and tentatively warranted by empirical laws, when the valid laws can be logically derived from the theories.

Like Hume they deny that either laws or theories can be permanently validated empirically, but they require that the general laws be founded in observation language as a condition for the objectivity needed for valid science. And they maintain that particularly quantified observation statements describing singular events are incorrigible and beyond revision.

All positivists reject the romantics' *verstehen* thesis of criticism. They argue that empathy is not a reliable tool, and that the methods of obtaining knowledge in the social sciences are the same as those used in the physical sciences. They complain that subjective *verstehen* may easily involve erroneous imputation of the idiosyncrasies of the observer's experiences or fantasies to the subjects of inquiry.

Explanation

Positivists and specifically Carl Hempel (1905-1997) and Paul Oppenheim (1885-1977) in their "Logic of Explanation" in the journal *Philosophy of Science* (1948) advocate the "covering-law" schema for explanation.

According to the "covering-law" schema for explanation, statements describing observable individual events are explained if they are derived deductively from other observation-language statements describing observable individual events together with "covering", *i.e.*, universally quantified empirical laws.

This concept of explanation has also been called the "deductive-nomological model".

The neopositivists also maintained that theories explain laws, when the theories are premises from which the empirical laws are deductively derived as theorems. The deduction is enabled by the mediation of "correspondence rules" also called "bridge principles". Correspondence rules are sentences that relate the theoretical terms in an explaining theory to the observation terms in the explained empirical laws.

2.03 Pragmatism

We are now said to be in a "postpositivist' era in the history of Western philosophy, but this term merely says that positivism has been relegated to history; it says nothing of what has replaced it. What has emerged is a new coherent master narrative appropriately called "contemporary pragmatism" or "realistic neopragmatism", which was occasioned by reflection on quantum theory, and is currently the ascendant philosophy in American academia. Contemporary pragmatism is a general philosophy for all empirical sciences, both social and natural sciences.

Pragmatism has earlier versions in the classical pragmatists, notably those of Charles Peirce, William James (1842-1910) and John Dewey (1859-1952). Some theses in classical pragmatism such as the importance of belief have been carried forward into the new. In contemporary pragmatism belief is strategic, because it controls relativized semantics, which signifies and thus reveals a correspondingly relativized ontology that is realistic to the degree that a belief is empirically adequate. Especially important is Dewey's emphasis on participation and his pragmatic thesis that the logical distinctions and methods of scientific inquiry develop out of scientists' successful problem-solving processes.

The provenance of the contemporary realistic pragmatist philosophy of science is 1932 Nobel-laureate physicist Werner Heisenberg's (1901-1976) reflections on the language in his revolutionary quantum theory in microphysics. There have been various alternative semantics and thus ontologies proposed for the quantum theory. Most physicists today have accepted one that has ambiguously been called the "Copenhagen interpretation".

There are two versions of the Copenhagen interpretation. Contrary to the alternative "hidden variables" view of David Bohm (1917-1992), ***both*** Copenhagen versions assert a thesis called "duality". The duality thesis is that the wave and particle manifestations of the electron are two aspects of the same entity, as Heisenberg says in his *Physical Principles of the Quantum Theory* (1930), rather than two separate entities, as Bohm says.

1922 Nobel-laureate Niels Bohr (1885-1962), founder of the Copenhagen Institute for Physics, proposed a version called "complementarity". His version says that the mathematical equations of quantum theory must be viewed instrumentally instead of descriptively, because only ordinary discourse and its refinement in the language of classical physics are able to describe physical reality. Instrumentalism is the doctrine that scientific theories are not descriptions of reality, but are meaningless yet useful linguistic instruments that enable correct predictions.

The quantum theory says that the electron has both wave and particle properties, but in classical physics the semantics of the terms "wave" and "particle" are mutually exclusive – a wave is spread out in space while a particle is a concentrated point. Therefore Bohr maintained that description of the electron's duality as both "wave" and "particle" is an empirically indispensable semantic antilogy that he called "complementarity".

Heisenberg, a colleague of Bohr at the Copenhagen Institute, proposed his alternative version of the Copenhagen interpretation. His version also contains the idea of the wave-particle duality, but he said that the mathematical expression of the quantum theory is realistic and descriptive rather than merely instrumental. And since the equations describing both the wave and particle properties of the electron are mathematically consistent, he disliked Bohr's complementarity antilogy. Years later Yale University's Hanson, an advocate of the Copenhagen physics, said that Bohr maintained a "naïve epistemology".

Duality is a thesis in physics while complementarity is a thesis in philosophy of language. These two versions of the Copenhagen interpretation differ not in their physics, but in their philosophy of language. Bohr's philosophy is called a "naturalistic" view of semantics, which requires what in his *Atomic Physics and the Description of Nature* (1934) he called "forms of perception". Heisenberg's philosophy is called an "artifactual" or a "conventionalist" view of semantics, in which the equations of the quantum theory supply the linguistic context, which defines the concepts that the physicist uses for observation.

1921 Nobel-laureate physicist Albert Einstein (1879-1955) had influenced Heisenberg's philosophy of language, which has later been incorporated into the contemporary pragmatist philosophy of language. And consistent with his relativized semantics Heisenberg effectively practiced ontological relativity and maintained that the quantum reality exists as "*potentia*" prior to determination to a wave or particle by execution of a measurement operation. For Heisenberg indeterminacy is real.

The term "complementarity" has since acquired some conventionality to signify duality, and is now ambiguous as to the issue between Bohr and Heisenberg, since physicists typically disregard the linguistic issue.

For more about Heisenberg and quantum theory the reader is referred to **BOOKs II** and **IV** at the free web site www.philsci.com or in the e-book *Twentieth-Century Philosophy of Science*: *A History,* which is available from most Internet booksellers.

The distinctive linguistic philosophy of Einstein and especially Heisenberg as incorporated into the contemporary pragmatist philosophy of science can be summarized in three theses, which may be taken as basic principles in contemporary pragmatism:

Thesis I: Relativized semantics

Relativized semantics is meanings defined by the linguistic context consisting of universally quantified statements believed to be true.

The seminal work is "Quantum Mechanics and a Talk with Einstein (1925-1926)" in Heisenberg's *Physics and Beyond* (1971). There Heisenberg relates that in April 1925, when he presented his matrix-mechanics quantum physics to the prestigious Physics Colloquium at the University of Berlin, Einstein, who was in the assembly, afterward invited Heisenberg to chat in his home that evening. In their conversation Einstein said that he no longer accepts the positivist view of observation including such positivist ideas as operational definitions. Instead he issued the aphorism: "the theory decides what the physicist can observe".

The event was historic. Einstein's aphorism about observation contradicts the fundamental positivist thesis that there is a natural dichotomous separation between the semantics of observation language and that of theory language. Positivists believed that the objectivity of science requires that the vocabulary and semantics used for incorrigible observation must be uncontaminated by the vocabulary and semantics of speculative and provisional theory.

In the next Chapter titled "Fresh Fields (1926-1927)" in the same book Heisenberg reports that Einstein's 1925 discussion with him in Berlin had later occasioned his own reconsideration of observation. Heisenberg recognized that classical Newtonian physical theory had led him to conceptualize the observed track of the electron as continuous in the cloud chamber – an instrument for microphysical observation developed by 1927 Nobel-laureate C.T.R. Wilson (1869-1961) – and therefore to misconceive the electron as simultaneously having a definite position and momentum like all Newtonian bodies in motion.

Recalling Einstein's aphorism that the theory decides what the physicist can observe, Heisenberg reconsidered what is observed in the cloud chamber. He rephrased his question about the electron tracks in the cloud chamber using the concepts of the new quantum theory instead of those of the classical Newtonian theory. He therefore reports that he asked himself: Can the quantum mechanics represent the fact that an electron finds itself approximately in a given place and that it moves approximately at a given momentum? In answer to this newly formulated question he found that these approximations can be represented mathematically. He reports that he then developed this mathematical representation, which he called the "uncertainty relations", the historic contribution for which he received the Nobel Prize in 1932.

Later Hanson expressed Einstein's aphorism that the theory decides what the physicist can observe by saying observation is "theory-laden" and likewise Popper (1902-1994) by saying it is "theory-impregnated". Thus for pragmatists the semantics of all descriptive terms is determined by the linguistic context consisting of universally quantified statements believed to be true.

In his *Against Method* (1975, Ch. 2-3) Paul Feyerabend (1924-1994) also recognized employment of relativized semantics to create new observation language for discovery, and he called the practice **"counterinduction".** To understand counterinduction, it is necessary to understand the pragmatic concept of "theory": theories are universally quantified statements that are proposed for testing. Feyerabend found that Galileo (1564-1642) had practiced counterinduction in the *Dialogue Concerning the Two Chief World Systems* (1632), where Galileo reinterpreted apparently falsifying observations in common experience by using the concepts from the apparently falsified heliocentric theory instead of the concepts from the prevailing geocentric theory. Likewise Heisenberg had also practiced counterinduction to reconceptualize the perceived sense stimuli observed as the electron track in the cloud chamber by using quantum concepts instead of classical Newtonian concepts, and he then developed the indeterminacy relations.

Counterinduction is using the semantics of an apparently falsified theory to revise the test-design language that had supplied the semantics of the language describing the apparently falsifying observations, and thereby to produce new language for observation.

Like Einstein, contemporary pragmatists say that the theory decides what the scientist can observe. Thus semantics is relativized in the sense that the meanings of descriptive terms used for reporting observations are not just names or labels for phenomena, but rather are formed by the context in which they occur.

More specifically in "Five Milestones of Empiricism" in his *Theories and Things* (1981) the neopragmatist philosopher of language Willard van Quine (1908-2000) says that the meanings of words are abstractions from the truth conditions of the sentences that contain them, and that it was this recognition of the semantic primacy of sentences that give us contextual definition.

The defining context consists of universally quantified statements that proponents believe to be true. The significance is that the acceptance of a new theory superseding an earlier one and sharing some of the same descriptive terms produces a semantical change in the descriptive terms shared by the theories and their common observation reports. The change consists of replacement of some semantical component parts in the meanings of the terms in the old theory with some parts in the meanings of the terms in the new theory.

Thus Einstein for example changed the meanings of such terms as "space" and "time", which occur in both the Newtonian and relativity theories. And Heisenberg changed the meanings of the terms "wave" and "particle", such that they are no longer mutually exclusive. Feyerabend calls the semantical change due to the relative nature of semantics "meaning variance".

For more about Quine the reader is referred to **BOOK III** at the free web site www.philsci.com or in the e-book *Twentieth-Century Philosophy of Science: A History,* which is available from most Internet booksellers.

Thesis II: Empirical underdetermination

Empirical underdetermination refers to the limited ability of the semantics of language at any given time to signify reality.

Experience is replete with instances in which it is unclear as to whether or not a descriptive term may apply. **Measurement error** and **conceptual vagueness**, which can be reduced indefinitely but never completely eliminated, exemplify the omnipresent and ever-present empirical underdetermination of descriptive language that occasions observational ambiguity and theoretical pluralism. But no semantically interpreted syntax is completely underdetermined empirically such that it is utterly devoid of any associated ontology.

Einstein recognized that a plurality of alternative but empirically adequate theories could be consistent with the same observational description, a situation that he called "an embarrassment of riches". Additional context including law statements in improved test-design language contributes additional semantics to the observational description in the test designs, thus reducing while never completely eliminating empirical under-determination.

In his *Word and Object* (1960) Quine introduced the phrase "empirical underdetermination", and wrote that the positivists' "theoretical" terms are merely more empirically underdetermined than terms they called "observation" terms. Thus contrary to positivists the types of terms are not qualitatively different. There Quine also says that reference to ontology is "inscrutable"; reference to relativized ontology is as inscrutable as signification by semantics is empirically underdetermined.

Thesis III: Ontological relativity

Heisenberg is a realist. In his *Physics and Philosophy*: *The Revolution in Modern Science* (1958) he says that the transition from the possible to the actual that takes place with the act of observation involves the interaction of the electron with the measuring device and applies to the physical and not to the psychological act of observation. He affirms that quantum theory does not contain "genuine subjective features" in the sense that it introduces the mind of the physicist as a part of the atomic event. Heisenberg also disliked Bohr's view that the equations of quantum theory must be viewed instrumentally, *i.e.*, they do not describe reality.

He practiced ontological relativity. In his discussions about Einstein's special theory of relativity in *Physics and Philosophy* and in *Across the Frontiers* (1974) Heisenberg describes the "decisive step" in the development of special relativity. That step was Einstein's rejection of 1902 Nobel-laureate Hendrik Lorentz's (1853-1928) distinction between "apparent time" and "actual time" in the Lorentz-Fitzgerald contraction. Lorentz took the Newtonian concepts to describe real space and time. But in his relativity theory Einstein took Lorentz's "apparent time" as physically real time, while altogether rejecting the Newtonian concept of absolute time as real time. In other words the "decisive step" in Einstein's special theory of relativity consisted of Einstein's taking the relativity theory realistically, thus letting his relativity theory characterize the physically real, i.e., physical ontology.

Heisenberg imitated Einstein's practice of ontological relativity for making his version of the Copenhagen interpretation of quantum physics.

In "History of Quantum Theory" in his *Physics and Philosophy*: *The Revolution in Modern Science* Heisenberg describes his use of Einstein in his discovery experience for quantum theory. There he states that his development of the indeterminacy relations involved turning around a question: instead of asking himself how one can express in the Newtonian mathematical scheme a given experimental situation, he asked whether only such experimental situations can arise in nature as can be described in the formalism of his quantum mechanics. The new question is an ontological question with the answer supplied by his quantum theory.

Again in "Remarks on the Origin of the Relations of Uncertainty" in *The Uncertainty Principle and Foundations of Quantum Mechanics* (1977) Heisenberg explicitly states that a Newtonian path of the electron in the cloud chamber does not exist. And still again in "The Development of the Interpretation of the Quantum Theory" in 1945 Nobel-laureate Wolfgang Pauli's *Niels Bohr and the Development of Physics* (1955) Heisenberg says that he inverted the question of how to pass from an experimentally given situation to its mathematical representation. There he concludes that only those states that can be represented as vectors in Hilbert space can exist in nature and be realized experimentally. And he immediately adds that this conclusion has its prototype in Einstein's special theory of relativity, when Einstein had removed the difficulties of electrodynamics by saying that the apparent time of the Lorentz transformation is real time.

Like Heisenberg in 1926, the contemporary pragmatist philosophers today let the scientist rather than the philosopher decide ontological questions. And the scientist decides on the basis of empirical adequacy demonstrated in his empirically tested explanations. Many years later in his *Ontological Relativity* (1970) Quine called this thesis "ontological relativity", as it is known today.

Ontological relativity did not begin with Heisenberg much less with Quine. Copernicus (1473-1543) and Galileo practiced it when they both interpreted heliocentrism realistically thus accepting the ontology it describes – to the fateful chagrin of Pope Urban VIII (1568-1644). Heisenberg's Copenhagen interpretation still prevails in physics today. The contemporary pragmatist concepts of the four functional topics may now be summarized as follows:

Aim of science

The successful outcome (and thus the aim) of basic-science research is explanations made by developing theories that satisfy critically empirical tests, and that are thereby made scientific laws, which can function in scientific explanations and test designs.

Wherever possible the explanation should enable prediction of either future events or evidence of past events, since the laws make universal claims. And it is pragmatically beneficial furthermore for the explanation to enable control of explained nonlinguistic reality by applied science thus enabling new engineering technologies, new medical therapies and new social policies.

Discovery

Discovery is the construction of new and empirically more adequate theories.

The semantics of newly constructed theories reveal new perspectives and ontologies.

A discovery system produces a transition from an input-language state description containing currently available information to an output-language state description containing generated and tested new theories.

Contemporary pragmatism is consistent with computerized discovery systems, which aim to proceduralize and then to mechanize new theory construction, in order to advance contemporary science.

In the "Introduction" to his magisterial *Patterns of Discovery: An Inquiry into the Conceptual Foundations of Science* (1958), Hanson wrote that earlier philosophers of science like the positivists had mistakenly regarded as paradigms of inquiry finished systems like Newton's planetary mechanics instead of the unsettled, dynamic research sciences like contemporary microphysics. Hanson explains that the finished systems are <u>no longer research sciences</u>, although they were at one time. He states that distinctions applying to the finished systems ought to be suspect when transferred to research disciplines, and that such transferred distinctions afford an artificial account of the activities in which Kepler, Galileo and Newton were actually engaged. He thus maintains that ideas such as "theory", "hypothesis", "law", "causality" and "principle" if drawn from what he calls the finished "catalogue-sciences" found in undergraduate textbooks will ill prepare one for research science.

Both romantics and positivists define "theory" semantically, while contemporary pragmatists define "theory" pragmatically, i.e., by its function in basic-research science. Contemporary pragmatists define both theory and observation language pragmatically instead of semantically; the pragmatics of both theory and observation uses of language in basic science is empirical testing. Neopragmatists recognize the nontruth-functional hypothetical-conditional schema for expressing proposed theories.

The pragmatics of theory is empirical testing to discover laws.

For pragmatists "theory" is universally quantified language that is *proposed* for testing, and "test-design" is universally quantified language that is *presumed* for testing.

For pragmatists scientific laws are former theories that have been tested with nonfalsifying test outcomes.

Neopragmatists identify theory language pragmatically as universally quantified statements proposed for testing, but they individuate theories semantically. Two theory expressions are different theories either (1) if the expressions have different test designs so they address different subjects, or (2) if the expressions make contrary claims about the same subject as defined by the same test design.

Criticism

Criticism pertains to the criteria for the acceptance or rejection of theories. **The *only* criterion for scientific criticism that is acknowledged by the contemporary pragmatist is the *empirical criterion* operative in an empirical test or otherwise experientially. An empirical test is:**

(1) an effective decision procedure that can be schematized as a *modus tollens* logical deduction from a set of one or several logically related universally quantified theory statements expressible in a nontruth-functional hypothetical-conditional schema

(2) together with a particularly quantified antecedent description of the initial test conditions as defined in a test design

(3) that jointly conclude to a consequent particularly quantified description of a produced (predicted) test-outcome event

(4) that is compared with the observed test-outcome description.

Test-designs are universally quantified statements that are presumed for testing. Test designs characterize the subject of the test, and describe procedures for execution of the test. They also include universal statements that are semantical rules for the test-outcome statements, which are asserted with particular quantification, when the test design is executed and the test outcome is produced.

Observation language is particularly quantified test-design and test-outcome statements with their semantics defined in the universally quantified test-design language including the test outcome language.

Unlike positivists, pragmatists do not recognize any natural observation semantics. For believers in a theory, the theory language may also contribute to the observational semantics, but that semantical contribution cannot operate in reporting the test outcome without violating the test's contingency.

On the pragmatist thesis of relativized semantics and ontological relativity, semantics and ontologies can never trump the empirical criterion for criticism, because acceptance of ontologies is based upon empirical adequacy of a theory especially as demonstrated by repeated nonfalsifying empirical test outcomes. Thus contrary to romantics, pragmatists permit description of intersubjective mental states in social-science theories and explanations, but unlike many sociologists and economists pragmatists never require or employ such description as a criterion for criticism.

Pragmatists recognize the *modus tollens* falsifying argument for empirical testing of the theories.

Unlike the logical positivists, pragmatists do not recognize the truth-functional conditional logic for scientific criticism, because the logic of empirical testing is not truth-functional.

Explanation

A scientific explanation is:

(1) a discourse that can be schematized as a *modus ponens* logical deduction from a set of one or several universally

quantified law statements expressible in a nontruth-functional hypothetical-conditional schema

(2) together with a particularly quantified antecedent description of realized initial conditions

(3) that jointly conclude to a consequent particularly quantified description of the explained event.

Explanation describes the occurrence of individual events and conditions as caused by the occurrence of other described events and conditions related in law statements.

Pragmatists recognize the *modus ponens* nontruth-functional deductive logical argument.

The argument has the hypothetical-conditional schema that includes a set of one or several universally quantified law statements expressible in conditional form. Whenever possible the explanation is predictive due to the universality claim of the theory.

Laws may be said to be "explained" in the sense that a set of logically related laws may be arranged into an axiomatic deductive system with some laws derived from other laws. Such deductive systems of statements or equations integrate their semantics to convey an impression of coherence, and thereby confer psychological satisfaction and may also offer some pedagogical utility. But such systems add nothing to empirical adequacy or to realizing the aim of science. Such rearranging is what in his *Patterns of Discovery* Hanson called mere "catalogue science" as opposed to "research science". However if the conclusions are not established laws because they are not yet untested, then they are not explanations, but they may be interesting new theories proposed for testing.

Chapter 3. Philosophy of Language

Basic scientific research produces language such as theories, test designs, observation reports, laws and explanations. Therefore many and probably most of the central concepts and issues in philosophy of science involve philosophy of language. Accordingly the following selected elements of contemporary pragmatist philosophy of language are here discussed in relation to philosophy of science.

3.01 Synchronic and Diachronic Analyses

To borrow some terminology from Ferdinand de Saussure's (1857-1913) classic *Course in General Linguistics* (1959) language analyses may be either **synchronic** or **diachronic.**

The synchronic view is static, because it exhibits the state of a language at a point in time like a photograph.

And to borrow some terminology from Carnap's *Meaning and Necessity* (1947) with modified meaning for computational philosophy of science, the state of the language for a specific scientific problem is displayed synchronically in a **semantical state description**. In the pragmatist's semantical state description statements including both the law language in the relevant test design and the theory language function as semantical rules that describe the meanings of their constituent descriptive terms.

The diachronic view on the other hand exhibits two chronologically successive states of the language for the same problem as defined by a single test design, and it shows semantical change over the interim period. Then the

view is a comparative-static semantical analysis like "before" and "after" photographs.

And if a transitional process between the two successive language states is also described, as in the computer code for a discovery system, then the diachronic view is dynamic like a motion picture film.

For more about Carnap the reader is referred to **BOOK III** at the free web site www.philsci.com or in the e-book *Twentieth-Century Philosophy of Science*: *A History,* which is available from most Internet booksellers.

3.02 Object Language and Metalanguage

Many philosophers of science such as Carnap in his *Logical Syntax of Language* (1937) distinguish two levels of language, **object language** and **metalanguage.**

Object language is used to describe the nonlinguistic real world.

Metalanguage is used to describe language, either object language or metalanguage.

The language of science is typically expressed in the object-language perspective, while much of the discourse in philosophy of science is in the metalinguistic perspective. Terms such as "theory", "law" and "explanation" are examples of expressions in metalanguage.

3.03 Dimensions of Language

The metalinguistic perspective includes what Carnap called dimensions of language, which serve well as an organizing framework for philosophy of language. Four dimensions may be distinguished for philosophy of language. They are A. **syntax,** B. **semantics,** C. **ontology** and D. **pragmatics.**

Syntax is the structure of language, semantics is the meanings associated with syntax, ontology is the real world as described by semantics, and pragmatics is the uses of semantically interpreted syntax.

Most philosophers of science ignore the linguists' phonetic and phonemic dimensions. And most linguists ignore the ontological dimension.

A. SYNTAX

3.04 Syntactical Dimension

Syntax is the system of linguistic symbols considered in abstraction from their associated meanings.

Syntax is the most obvious part of language. It is residual after abstraction from pragmatics, ontology, and semantics. And it consists only of the forms of expressions, so it is often said to be "formal". Since meanings are excluded from the syntactical dimension, the expressions are said to be "semantically uninterpreted". And since much of the language of science is usually written, the syntax of interest consists of visible marks on paper or more recently linguistic source-code displays on computer monitor display screens. The syntax of expressions is also sometimes called "inscriptions". Examples of syntax include the sentence structures of colloquial discourse, the formulas of pure or formal mathematics, and computer source codes such as **FORTRAN** or **LISP.**

3.05 Syntactical Rules

Syntax is a *system* of symbols. Therefore in addition to the syntactical symbols and structures, there are also rules for the system called "syntactical rules". These rules are of two types: **formation rules** and **transformation rules.**

Formation rules are procedures described in metalanguage that regulate the construction of grammatical expressions out of more elementary symbols, usually terms.

A generative grammar applies formation rules to produce grammatical expressions from inputs consisting of terms.

A mechanized generative grammar is a computer system that implements a generative grammar.

A discovery system is a mechanized generative grammar that constructs and usually also empirically tests new scientific theories as its output.

Formation rules order such syntactical elements as mathematical variables and operator signs, descriptive and syncategorematic terms in logic, and the user-defined variable names and reserved words in computer source codes. Expressions constructed from these symbols in compliance with the formation rules for a language are called "grammatical" sentences or "well formed formulas", and include the computer instructions called "compiler-acceptable" and "interpreter-acceptable" source code.

When there exists an explicit and adequate set of syntactical formation rules, it is possible to develop a type of computer program called a "mechanized generative grammar".

The mechanized generative-grammar computer programs input, process, and output object language, while the source-code instructions constituting the computer system are therefore metalinguistic expressions.

When a mechanized generative grammar is used to produce new scientific theories in the object language of a science, the computer system is what Simon called a "**discovery system**". Typically the system also contains an empirical test criterion to select for output a subset of the deluge of theories generated.

Transformation rules change grammatical sentences into other grammatical sentences. Transformation rules are used in logical and mathematical deductions. But logic and mathematical rules are intended not only to produce new grammatical sentences but also to guarantee truth transferability from one set of sentences or equations to another to generate theorems, usually by the transformation rule of substitution that makes mathematics and logic extensional.

Today except for the logic of testing and explanation transformation rules are of greater interest to linguists, mathematicians, logicians and scientists than to contemporary philosophers of science, who recently have been more interested in mechanizing formation rules for generative-grammar discovery systems.

In 1956 Simon developed a computer system named **LOGIC THEORIST**, which operated with his "heuristic-search" discovery system design. This system developed deductive proofs of the theorems in Alfred N. Whitehead (1861-1947) and Bertrand Russell's (1872-1970) *Principia Mathematica* (1910-1913). The symbolic-logic formulas are object language for this system. Simon correctly denies that the

Russellian symbolic logic is an effective metalanguage for the design of discovery systems.

3.06 Mathematical Language

The syntactical dimension of mathematical language includes mathematical symbols and the formation and transformation rules of the various branches of mathematics. Mathematics applied in science functions as object language for which the syntax is supplied by the mathematical formalism. Often the object language of science is mathematical rather than colloquial, because measurement values for descriptive variables enable the scientist to quantify the error in his theory after estimates are made for the range of inevitable measurement error usually estimated by repeated execution of the measurement procedure.

3.07 Logical Quantification in Mathematics

Mathematical expressions in science are universally quantified when descriptive variables have no associated numerical values, and are particularly quantified when numeric values are associated with the expression's descriptive variables either by measurement or by calculation.

Like categorical statements, mathematical equations are explicitly quantified logically as either universal or particular, even though the explicit indication is not by means of the syncategorematic logical quantifiers "every", "some" or "no". An equation in science is universally quantified logically when none of its descriptive variables are assigned numeric values. Universally quantified equations may also contain mathematical descriptive constants as in some theories or laws. An equation is particularly quantified logically by associating measurement values with any of its descriptive variables. A variable may then be said to describe an **individual measurement instance.**

When a numeric value is associated with a descriptive variable by computation with measurement values associated with other descriptive variables in the same mathematical expression, then the variable's calculated value may be said to describe an **individual empirical instance**. In this case the referenced instance has not been measured but depends on measurements associated with other variables in the same equation.

Individual empirical instances are calculated when an equation is used to make a numerical prediction. The individual empirical instance is the predicted value, which makes an empirical claim. In a test it is compared with an individual measurement instance, which is the test-outcome value made for the same variable. The individual empirical instance made by the predicting equation is not said to be empirical because the predicting equation is known to be correct or accurate, but rather because the predicting theory makes an empirical claim that may be falsified by the empirical test, when the predicted empirical instance falls outside the range of estimated measurement error in the individual measurement instance for the test-outcome value for the same variable.

B. SEMANTICS

3.08 Semantical Dimension

Semantics is the meanings associated with syntactical symbols.

Semantics is the second of the four dimensions, and it includes the syntactical dimension. Language viewed in the semantical metalinguistic perspective is said to be "semantically interpreted syntax", which is merely to say that the syntactical symbols have meanings associated with them.

3.09 Nominalist vs. Conceptualist Semantics

Both nominalism and conceptualism are represented in contemporary pragmatism, but nominalism is the minority view. Contemporary nominalist philosophers advocate a two-level semantics, which in written language consists only of syntactical structures and the ontologies that are referenced by the structures, or as Quine says "word and object". The two-level semantics is also called a referential thesis of semantics, because it excludes any mid-level mental representations variously called ideas, meanings, significations, concepts or propositions. Therefore on the nominalist view language purporting to reference nonexistent fictional entities is semantically nonsignificant, which is to say it is literally meaningless.

On the alternative three-level view terms symbolize universal meanings, which in turn signify such aspects of extramental reality as attributes, and reference ontologies that include individual entities. When we are exposed to the extramental realities, they are distinguishable by the senses in perceived stimuli, which in turn are synthesized by the brain, and may then be registered in memory. The sense stimuli deliver information revealing similarities and differences in reality. The signified attributes are similarities found by perception, and the referenced entities manifesting the attributes are recognized by invariant continuities found in perceived change. The signification is always more or less vague, and the reference is therefore always more or less indeterminate, and the ontology is therefore more or less what Quine calls "inscrutable". The three-level view is often called a conceptualist thesis of semantics.

The philosophy of nominalism was common among many positivists, although some like the logical positivist Carnap maintained a three-level semantics. In Carnap's three-level semantics descriptive terms symbolize what he called "intensions", which are concepts or meanings effectively viewed as in simple supposition (See below, Section **3.26**). The intensions in turn signify attributes and thereby reference in personal supposition what Carnap called "extensions", which are the individual entities manifesting the signified attributes.

While the contemporary pragmatism emerged as a critique of neopositivism, some philosophers carried the positivists' nominalism into contemporary pragmatism. A few pragmatist philosophers such as Quine opted for nominalism. He rejected concepts, ideas, meanings, propositions and all other mentalistic views of knowledge due to his acceptance of the notational conventions of the Russellian first-order predicate calculus, a logic that Quine liked to call "canonical". However, in his book *Word and Object* (1960) Quine also uses a phrase "stimulus meaning", which he defines as a disposition by a native speaker of a language to assent or dissent from a sentence in response to present stimuli. And he added that the stimulus is not just a singular event, but rather is a "universal", which he called a "repeatable event form".

Nominalism is by no means essential to or characteristic of contemporary pragmatism, and most contemporary pragmatist philosophers of science such as Hanson, Feyerabend and Thomas S. Kuhn (1922-1996), and most linguists except the behaviorists have opted for the three-level semantics, which is also assumed herein. Behaviorism is positivism in the behavioral sciences. Also, computational philosophers of science such as Simon, Langley and Thagard, who advocate the cognitive-psychology interpretation of discovery systems instead of the linguistic-analysis interpretation, also reject both nominalism and behaviorism (See below, Section **3.34**).

Cognitive scientists recognize the three-level semantics, and furthermore believe that they can model the mental level with computer systems. Thus in his book *Mind: Introduction to Cognitive Science* (1996) Thagard states that the central hypothesis of cognitive science is that the human mind has mental representations analogous to data structures and cognitive processes analogous to algorithms. Cognitive psychologists claim that their computer systems using data structures and algorithms applied to the data structures can model both the mind's concepts and its cognitive processes with the concepts.

3.10 Naturalistic vs. Artifactual Semantics

The artifactual thesis of the semantics of language is that the meanings of descriptive term are determined by their linguistic context consisting of universally quantified statements believed to be true.

The contemporary pragmatist philosophy of science is distinguished by a post-positivist philosophy of language, which has replaced the traditional naturalistic thesis with the artifactual thesis of semantics. The artifactual thesis implies that ontology, semantics and belief are mutually determining.

The naturalistic thesis affirms an absolutist semantics according to which the semantics of descriptive terms is passively acquired ostensively, and is fully determined by perceived reality and the processes of perception.

Thus on the naturalistic view descriptive terms function effectively as names or labels, a view that Quine ridicules with his phrases "myth of the museum" and "gallery of ideas". Then after the meanings for descriptive terms are acquired ostensively, the truth of statements constructed with the descriptive terms is ascertained empirically.

On the artifactual semantical thesis sense stimuli reveal mind-independent reality as semantically signified ontology. Sense stimuli are conceptualized as the semantics that is formed by linguistic context consisting of a set of beliefs that by virtue of the set's belief status has a defining rôle for the semantics. When the beliefs that are laws function as test-design statements for a theory, the tests may occasion falsification of the proposed theory.

The artifactual semantical thesis together with the ontological relativity thesis revolutionized philosophy of science by subordinating both semantics and ontology to belief, especially empirically warranted belief. The outcome of this new linguistic philosophy is that ontology, semantics and belief are all mutually determining and thus interdependent.

3.11 Romantic Semantics

For romantics the semantics for social sciences explaining human action *must* include description of the culturally shared intersubjective meanings and consequent motivations that the human action has for the members of a social group.

On the romantic view the positivist semantics may be acceptable for the natural sciences, but it is deemed inadequate for understanding "human action" in the sociocultural sciences. "Human action" considered by the romantic cultural sciences has intersubjective meaning for the members of a group or society, and it is purposeful and motivating for the members' social interactions.

Romantics call the resulting intersubjective semantics "interpretative understanding". The social member's voluntary actions are controlled by this interpretative understanding, *i.e.*, by the motivating views and values that are internalized and shared among the members of a social group by the social-psychological "mechanism" of socialization, and are reinforced by the social psychological "mechanism" of social control. This understanding is accessed by the social scientist in the process of his research. Furthermore if the researcher is a member in the society or group he is investigating, the validity of his empathetically based and vicariously imputed interpretative understanding is deemed enhanced by his personal experiences as a participant in the group or society's life.

3.12 Positivist Semantics

For positivists the semantics of observation language is causally determined by nature and acquired ostensively by perception. Positivists maintain the naturalistic philosophy of semantics, and the semantics for descriptive terms used for reporting observations are primitive and simple.

All positivists maintain a naturalistic philosophy of semantics. These meanings were variously called "sensations", "sense impressions", "sense perceptions", "sense data" or "phenomena" by different positivists. For these often called "phenomenalists" the sense perceptions are the object of knowledge rather than constituting knowledge thus making positivism solipsistic.

Consider the case of a term such as "black": The child's ostensive acquisition of meaning might involve the child pointing his finger at a present instance of a perceived black object. And then upon hearing the word "black" in repeated experiences of several other black objects, he associates the word "black" with his various experienced perceptions of the color black. Furthermore from several early experiences expressible as "That crow is black" the young learner may eventually infer intuitively by natural inductive generalization that "Every crow is black." However, solipsistic phenomenalism makes the intersubjective sharing of such experiences philosophically problematic.

Positivists maintain three characteristic theses about semantics:
- Meaning invariance.
- Analytic-synthetic dichotomy.
- Observation-theory dichotomy.

3.13 Positivist Thesis of Meaning Invariance

The naturalistic semantics thesis implies that the semantics of a univocal descriptive term used to report observations is invariant through time and is independent of different linguistic contexts in which the term may occur.

What is fundamental to the naturalistic philosophy of semantics is the thesis that the semantics of observation terms is fully determined by the ostensive awareness that is perception, such that the resulting semantics is primitive and simple. Different languages are conventional in their vocabulary symbols and in their syntactical structures and grammatical rules. But according to the naturalistic philosophy of semantics nature makes the semantics of observation terms the same for all persons who have received the same perceptual stimuli that occasioned their having acquired their semantics in the same circumstances by simple ostension.

Positivists viewed this meaning invariance as the basis for objectivity in science.

3.14 Positivist Analytic-Synthetic Dichotomy

In addition to the descriptive observation terms that have primitive and simple semantics acquired ostensively, the positivist philosophers also recognized the existence of certain terms that acquire their meanings contextually and that have complex semantics. An early distinction between simple and complex ideas can be found in his *Essay Concerning Human Understanding* (1690) by the British empiricist philosopher John Locke (1632-1704). The positivist recognized compositional meanings for terms occurring in three types of statements: the **definition,** the **analytic sentence** and the **theory:**

The *first type* of term having complex semantics that the positivists recognized occurs in the definition. The defined subject term or *definiendum* has a compositional semantics that is exhibited by the structured meaning complex associated with the several words in the defining predicate or *definiens*. For example "Every bachelor is a never-married man" is a definition, so the component parts of the word "bachelor" are "never-married" and "man".

The *second type* occurs in the analytic sentence, which is an a priori or self-evident truth, a truth known by reflection on the interdependence of the meanings of its constituent terms. Analytic sentences contrast with synthetic sentences, which are a posteriori, *i.e.*, empirical and have independent meanings for their terms. The positivists view the analytic-synthetic distinction as a fundamental dichotomy between two types of statements. A similar distinction between "relations of ideas" and "matters of fact" can be found in *An*

Enquiry Concerning Human Understanding (1748) by the British empiricist philosopher David Hume.

An example of an analytic sentence is "Every bachelor is unmarried". The semantics of the term "bachelor" is compositional and is determined contextually, because the idea of never having been married is by definition included as a component part of the meaning of "bachelor" thus making the phrase "unmarried bachelor" redundant. Contemporary pragmatists such as Quine in his historic paper "Two Dogmas of Empiricism" (1952) reject the positivist thesis of a priori truth. Quine, who is a pragmatist, maintains that all sentences are actually empirical.

3.15 Positivist Observation-Theory Dichotomy

Positivists alleged the existence of "observation terms", which are terms that reference observable entities or phenomena. Observation terms are deemed to have simple, elementary and primitive semantics and to receive their semantics ostensively and passively. Positivists furthermore called the particularly quantified sentences containing only such terms "observation sentences", if stated on the occasion of observing. For example the sentence "That crow is black" uttered while the speaker of the sentence is viewing a present crow, is an observation sentence.

In contrast to observation terms there is a *third type* of term having complex semantics that the positivists called the "theoretical term". The term "electron" is a favorite paradigm for the positivists' theoretical term. The positivists considered theoretical entities such as electrons to be postulated entities as opposed to observed entities like elephants. And they defined "theory" as universally quantified sentences containing any theoretical terms. Many positivists view the semantics of the meaningful theoretical term to be simple like the observation term even though its semantics is not acquired by observation but rather contextually. Carnap was a more sophisticated positivist. He said that the definition determines the whole meaning of a defined term, while the theory determines only part of the meaning of a theoretical term, such that the theoretical term can acquire more meaning as the science containing the theory is developed.

Nominalists furthermore believe that theoretical terms are meaningless, unless these terms logically derive their semantics from observation terms. On the nominalists' view terms purporting either unobserved entities or phenomena not known observationally to exist have no known referents and therefore no semantical significance or meaning. For example the phrase "tooth fairy" is literally meaningless, since tooth fairies are deemed mythical and thus never to have been observed. For nominalists theoretical terms in science receive their semantics by logical connection to observation language by "correspondence rules", a connection that enables what positivists called "logical reduction to an observation-language reduction base". Without such connection the theory is deemed to be meaningless and objectionably "metaphysical".

Both the post-positivist Popper and later the formerly logical positivist Carl Hempel (1905-1997) have noted that the problem of the logical reduction of theories to observation language is a problem that the positivists have never solved, because positivists cannot exclude what they considered to be metaphysical and thus meaningless discourse from the scientific theories currently accepted by the neopositivists as well as by contemporary scientists. This unsolvable problem made the positivists' observation/theory dichotomy futile.

In summary the positivists recognized the definition, the analytical sentence and the theory sentence as exhibiting composition in the semantics of their constituent subject terms.

3.16 Contemporary Pragmatist Semantics

Heisenberg's reflection on the development of quantum physics anticipated development of the contemporary pragmatist philosophy thus initiating the New Pragmatism.

A fundamental postulate in the contemporary pragmatist philosophy of language is the rejection of the naturalistic thesis of the semantics of language and its replacement with the artifactual thesis that relativizes all semantics and ontology to linguistic context consisting of universally quantified beliefs.

The rejection of the naturalistic thesis is not new in linguistics, but it is fundamentally opposed to the positivism that preceded contemporary pragmatism.

3.17 Pragmatist Semantics Illustrated

Consider the following analogy illustrating relativized semantics. Our linguistic system can be viewed as analogous to a mathematical simultaneous-equation system. The equations of the system are a constraining context that determines the variables' numerical values constituting a solution set for the equation system. If there is not a sufficient number of constraining equations, the system is mathematically underdetermined such that there is an indefinitely large number of possible numerical solution sets.

In pure mathematics numerical underdetermination can be eliminated and the system can be made uniquely determinate by adding related independent equations, until there are just as many equations as there are variables. Then there is one uniquely determined solution set of numerical values for the equation system.

When applying such a mathematically uniquely determined equation system to reality as in basic science or in engineering, the pure mathematics functions as the syntax for a descriptive language, when the numerical values of the descriptive variables are measurements. But the measurement values make the mathematically uniquely determined equation system **empirically underdetermined** due to measurement errors, which can be reduced indefinitely but never completely eliminated. Then even for a mathematically uniquely determined equation system admitting only one solution set of numerical values, there are still an indefinitely large number of possible measurement values falling within even a narrow range of empirical underdetermination due to inevitable measurement errors.

When the simultaneous system of equations expresses an empirical theory in a test, and if its uniquely determined solution-set numerical values fall within the estimated range of measurement error in the corresponding measurement values produced in a test, then the theory is deemed not falsified. But if any of the uniquely determined solution-set numerical values are outside the estimated range of measurement error in the measurement values, then the theory is deemed to have been falsified by all who accept the falsifying test design and its execution.

Our descriptive language system is like a mathematically underdetermined system of equations having an indefinitely large number of solution sets for the system. A set of logically consistent beliefs constituting a system of universally quantified related statements is a constraining context that determines the semantics of the descriptive terms in the belief system. This is most evident in but not unique to an axiomatized deductive system. Like the equation system's numerical values the language system's semantics for any "semantical solution set", as it were, are relativized to one another by the system's universally quantified beliefs that have definitional force. But the semantics conceptualizing sense stimuli always contains residual vagueness. Due to this vagueness the linguistic system is empirically underdetermined and admits to an indefinitely large number of relativized semantical solution sets for the system. Unlike pure mathematics there never exists a uniquely determinate belief system of concepts.

This vagueness does not routinely manifest itself or cause communication problems and is deceptively obscured, so long as we encounter expected or familiar experiences for which our conventionalized beliefs are prepared. But the language user may on occasion encounter a new situation, which the existing relevant conventional beliefs cannot take into account. In such new situations the language user must make some decisions about the applicability of one or several of the problematic terms in his existing beliefs, and then add some new clarifying beliefs, if the decision about applicability is not *ad hoc*. This is true even of terms describing quantized objects.

Adding more universally quantified statements to the belief system reduces this empirical underdetermination by adding clarifying information, but the residual vagueness can never be completely eliminated. Our semantics captures determinate mind-independent reality, but the cognitive capture with our semantics can never be exhaustive. There is always residual vagueness in our semantics. Vagueness and measurement error are both manifestations of empirical underdetermination. And increased clarity reduces vagueness as increased accuracy reduces measurement error.

Relativized semantics also has implications for ontology. Mind-independent recalcitrant reality imposes the empirical constraint that makes our belief systems contingent, and in due course falsifies them. Our access to mind-independent reality is by language-dependent relativized semantics, which signifies a corresponding ontology. Ontology is the cognitively apprehended aspects of the fathomless plenitude that is mind-independent reality as described by the relativized semantics. Thus there are no referentially absolute or fixed terms. Instead descriptive terms are always referentially fuzzy or as Quine says "inscrutable", because their semantics is always empirically underdetermined.

There are three noteworthy consequences of the artifactual thesis of relativized semantics:

-Rejection of the positivist observation-theory dichotomy.
-Rejection of the positivist thesis of meaning invariance.
-Rejection of the positivist analytic-synthetic dichotomy.

3.18 Rejection of the Observation-Theory Dichotomy

All descriptive terms are empirically underdetermined, such that what the positivists called "theoretical terms" are simply descriptive terms that are more empirically underdetermined than what the positivists called "observation terms".

One of the motivations for the positivists' accepting the observation-theory dichotomy is the survival of the ancient belief that science in one respect or another has some permanent and incorrigible foundation that distinguishes it as true knowledge as opposed to mere speculation or opinion. In the positivists' foundational agenda observational description is presumed to deliver this certitude, while theory language is subject to revision, which is sometimes revolutionary in scope. The positivists were among the last to believe in any such eternal verities as the defining characteristic of truly scientific knowledge.

More than a quarter of a century after Einstein told Heisenberg that the theory decides what the physicist can observe, and after Heisenberg said he could observe the electron in the cloud chamber, philosophers of science began to reconsider the concept of observation, a concept that had previously seemed inherently obvious. On the contemporary pragmatist view there are no observation terms that receive isolated meanings merely by simple ostension, and there is no distinctive or natural semantics for identifying language used for observational reporting. Instead every descriptive term is embedded in an interconnected system of beliefs, which Quine calls the "web of belief". A small relevant subset of the totality of beliefs constitutes a context for determining any given descriptive term's meaning, and a unilingual dictionary's relevant lexical entries are a minimal listing of a subset of relevant beliefs for each univocal term. Thus the positivists' thesis of "observation terms" is rejected by pragmatists.

All descriptive terms lie on a spectrum of greater or lesser empirical underdetermination. Contemporary pragmatists view the positivist problem of the reduction of theoretical terms to observation terms as a pseudo problem, or what Heisenberg called a "false question", and they view both observation terms and theoretical terms as positivist fabrications.

3.19 Rejection of Meaning Invariance

The semantics of every descriptive term is determined by the term's linguistic context consisting of a set of universally quantified statements believed to be true, such that a change in any of those contextual beliefs

changes some component parts of the constituent terms' meanings.

In science the linguistic context consisting of universally quantified statements believed to be true may include both theories undergoing or awaiting empirical testing and law statements used in test designs, which jointly contribute to the semantics of their shared constituent descriptive terms.

When the observation-theory dichotomy is rejected, the language that reports observations becomes subject to semantical change or what Feyerabend called "meaning variance". For the convinced believer in a theory the statements of the theory contribute meaning parts to the semantics of descriptive language used to report observations, such that for the believer a revision of the theory changes part of the semantics of the relevant observational description.

3.20 Rejection of the Analytic-Synthetic Dichotomy

All universally quantified affirmations believed to be true are both analytic and empirical.

On the positivist view the truth of analytic sentences can be known a priori, i.e., by reflection on the meanings of the constituent descriptive terms, while synthetic sentences require empirical investigation to determine their truth status, such that their truth can only be known a posteriori. Thus to know the truth status of the analytic sentence "Every bachelor is unmarried", it is unnecessary to take a survey of bachelors to determine whether or not any such men are currently married. However, determining the truth status of the sentence "Every crow is black" requires an empirical investigation of the crow-bird population and then a generalizing inference.

On the alternative neopragmatist view the semantics of all descriptive terms are contextually determined, such that all universally quantified affirmations believed to be true are analytic statements. But their truth status is not thereby known a priori, because they are also synthetic, *i.e.*, empirical, firstly known a posteriori by experience.

This dualism implies that when any universally quantified affirmation is believed to be empirically true, the sentence can then be used analytically, such that the meaning of its predicate offers a partial analysis of the meaning of its subject term. To express this analytic-empirical dualism Quine used the phrase "analytical hypotheses".

Thus "Every crow is black" is as analytic as "Every bachelor is unmarried", so long as both statements are believed to be true. The meaning of "bachelor" includes the idea of being unmarried and makes the phrase "unmarried bachelor" pleonastic. Similarly so long as one believes that all crows are black, then the meaning of "crow" includes the idea of being black and makes the phrase "black crow" pleonastic. The only difference between the beliefs is the degree of conventionality in usage, such that the phrase "married bachelor" seems more antilogous than the phrase "white crow". In science the most important reason for belief is empirical adequacy demonstrated by reproducible and repeated nonfalsifying empirical test outcomes.

3.21 Semantical Rules

A semantical rule is a universally quantified affirmation believed to be true and viewed in logical supposition in the metalinguistic perspective, such that the meaning of the predicate term displays some of the <u>component parts</u> of the meaning of the subject term.

The above discussion of analyticity leads immediately to the idea of "semantical rules", a phrase also found in the writings of such philosophers as Carnap and Alonzo Church (1903-1995) but with a different meaning in computational philosophy of science in the pragmatist context. In the contemporary pragmatist philosophy semantical rules are statements in the metalinguistic perspective, because they are about language. And their constituent terms are viewed in logical supposition, because as semantical rules the statements are about meanings as opposed to nonlinguistic reality (See below, Section **3.26**).

Semantical rules are enabled by the complex nature of the semantics of descriptive terms. But due to psychological habit that enables prereflective linguistic fluency, meanings are experienced wholistically and unreflectively. Thus if a fluent speaker of English were asked about crows, his answer would likely be in ontological terms such as the real creature's black color rather than as a reflection on the componential semantics of the term "crow" with its semantical component of black. Reflective semantical analysis is needed to appreciate the componential nature of the meanings of descriptive terms.

In scientific language the context consisting of universally quantified statements believed to be true includes both law statements in test-designs and theories for testing, which jointly contribute meaning components to the semantics of their shared descriptive terms.

3.22 Componential vs. Wholistic Semantics

On the neopragmatist view when there is a transition from an old theory to a new theory having the same test design, for the advocates of the old theory there occurs a semantical change in the descriptive terms shared by the old and new theories, due to the replacement of some

of the meaning parts of the old theory with meaning parts from the new theory. But the meaning parts contributed by the common test-design language remain unaffected.

Semantical change had vexed the contemporary pragmatists, when they initially accepted the artifactual thesis of the semantics of language. When they rejected a priori analytic truth, many of them mistakenly also rejected analyticity altogether. And when they accepted the contextual determination of meaning, they mistakenly took an indefinitely large context as the elemental unit of language for consideration. They typically construed this elemental context as consisting of either an explicitly stated whole theory with no criteria for individuating theories, or an even more inclusive "paradigm", i.e., a whole theory together with many associated pre-articulate skills and tacit beliefs. This wholistic (or "holistic") semantical thesis is due to using the psychological experience of meaning instead of making semantic analyses that enable recognition of the componential nature of lexical meaning.

On this wholistic view therefore a new theory that succeeds an alternative older one must, as Feyerabend maintains, completely replace the older theory including all its observational semantics and ontology, because its semantics is viewed as an indivisible unit. In his *Patterns of Discovery* Hanson attempted to explain such wholism in terms of Gestalt psychology. And following Hanson the historian of science Kuhn, who wrote a popular monograph titled *Structure of Scientific Revolutions*, explained the complete replacement of an old theory by a newer one as a "Gestalt switch".

Feyerabend tenaciously maintained wholism, but attempted to explain it by his own interpretation of an ambiguity he found in Benjamin Lee Whorf's (1897-1941) thesis of linguistic relativity also known as the "Sapir-Whorf hypothesis" formulated jointly by Whorf and Edward Sapir (1884-1931), a Yale University Linguist. In his "Explanation, Reduction and Empiricism", in *Minnesota Studies in the Philosophy of Science* (1962) and in his *Against Method* Feyerabend proposes semantic "incommensurability", which he says is evident when an alternative theory is not recognized to be an alternative. He cites the transition from classical to quantum physics as an example of such incommensurability.

The thesis of semantic incommensurability was also advocated by Kuhn, who proposed "incommensurability with comparability". But incommensurability with comparability is inconsistent as Hilary Putnam (1926-) observed in his *Reason, Truth and History* (1981), because comparison presupposes that there are some commensurabilities. Kuhn then revised the idea to admit "partial" incommensurability that he believed enables incommensurability with comparability but without successfully explaining how incommensurability can be partial.

Semantic incommensurability can only occur in language describing phenomena that have never previously been observed, *i.e.,* an observation for which the current state of the language has no semantic values (See below, Section **3.24**). But it is very seldom that a new observation is indescribable with the stock of existing descriptive terms in the language. The scientist may be able to resort to what Hanson called "phenomenal seeing". Furthermore incommensurability does not occur in theory revision, which is a reorganization of pre-existing descriptive language.

A wholistic semantical thesis including notably the semantic incommensurability thesis creates a pseudo problem for the decidability of empirical testing in science, when it implies complete replacement of the semantics of the descriptive terms used for test design and observation. Complete replacement deprives the two alternative theories of any semantical continuity, such that their language cannot even describe the same phenomena or address the same problem. In fact the new theory cannot even be said to be an alternative to the old one, much less an empirically more adequate one. Such empirical undecidability due to alleged semantical wholism would logically deny science both production of progress and recognition of its history of advancement. The untenable character of the situation is comparable to the French entomologist August Magnan whose book titled *Insect Flight* (1934) set forth a contemporary aerodynamic analysis proving that bees cannot fly. But bees in fact do fly, and empirical tests in fact do decide.

The thesis of componential semantics resolves the wholistic semantical muddle in the linguistic theses proffered by philosophers such as Hanson, Kuhn and Feyerabend. Philosophers of science have overlooked componential semantics, but linguists have long recognized componential analysis in semantics, as may be found for example in George L. Dillon's (1944) *Introduction to Contemporary Linguistic Semantics* (1977). Some other linguists use the phrase "lexical decomposition". With the componential semantical thesis it is unnecessary to accept any wholistic view of semantics much less any incommensurable discontinuity in language in theory development.

The expression of the componential aspect of semantics most familiar to philosophers of language is the analytic statement. But the neopragmatists' rejection of the analytic-synthetic dichotomy with its a priori truth claim need not imply the rejection of analyticity as such. The contextual determination of meaning exploits the analytic-empirical dualism. When there is a semantical change in the descriptive terms in a system of beliefs due to a revision of some of the beliefs, some component parts of the terms' complex meanings remain unaffected, while other parts are dropped and new ones added. Thus on the neopragmatist view when there is a transition from an old theory to a new theory having the same test design, for the advocates of the old theory there occurs a semantical change in the descriptive terms shared by the old and new theories, due to the replacement of the meaning parts from the old theory with meaning parts from the new theory, while the shared meaning parts contributed by the common test-design language remain unaffected.

For empirical testing in science the component meaning parts that remain unaffected by the change from one theory to a later alternative one consist of those parts contributed by the statements of test design shared by the two theories. Therein is found the semantical continuity that enables empirical testing of alternative theories to be decidable between them.

Thus a revolutionary change in scientific theory, such as the replacement of Newton's theory of gravitation with Einstein's, has the effect of changing only part of the semantics of the terms common to both the old and new theories. It leaves the semantics supplied by test-design language unaffected, so Arthur Eddington (1882-1942) could test both Newton's and Einstein's theories of gravitation simultaneously by describing the celestial photographic observations in his 1919-eclipse test. There is no semantic incommensurability between these theories.

For more about the philosophies of Kuhn, Feyerabend, and Eddington's 1919-eclipse test readers are referred to **BOOK VI** at the free web site www.philsci.com or in the e-book *Twentieth-Century Philosophy of Science*: *A History,* which is available from most Internet booksellers.

3.23 Componential Artifactual Semantics Illustrated

The set of affirmations believed to be true and predicating characteristics universally and univocally of the term “crow” such as “Every crow is black” are semantical rules describing component parts of the complex meaning of “crow”. But if a field ornithologist captures a white bird specimen that exhibits all the characteristics of a crow except its black color, he must make a semantical decision. He must decide whether he will continue to believe “Every crow is black” and that he holds in his birdcage some kind of white noncrow bird, or whether he will no longer believe “Every crow is black” and that the white bird in his birdcage is a white crow. Thus a semantical decision must be made. Color could be made a criterion for species identification instead of the ability to breed, although many other beliefs would also then be affected, an inconvenience that is typically avoided as a disturbing violation of the linguistic preference that Quine calls the principle of “minimum mutilation” of the web of belief.

Use of statements like “Every crow is black” may seem simplistic for science (if not quite bird-brained). But as it happens, a noteworthy revision in the semantics and ontology of birds has occurred due to a five-year genetic study launched by the Field Museum of Natural History in Chicago, the results of which were reported in the journal *Science* in June 2008. An extensive computer analysis of 30,000 pieces of nineteen bird genes showed that contrary to previously held belief falcons are genetically more closely related to parrots than to hawks, and furthermore that falcons should no longer be classified in the biological order originally named for them. As a result of the new genetic basis for classification, the American Ornithologists Union has revised its official organization of bird species, and many bird watchers’ field guides have been revised accordingly. Now well informed bird watchers will classify, conceptualize and observe falcons differently, because some parts of the meaning complex for the term "falcon" have been replaced with a genetically based conceptualization. Yet given the complexity of genetics some biologists argue that the concept of species is arbitrary.

Our semantical decisions alone neither create, nor annihilate, nor change mind-independent reality. But semantical decisions may change our mind-dependent linguistic characterizations of mind-independent reality and thus the ontologies, i.e., the various aspects of reality that the changed semantics reveals. This is due to the perspectivist nature of relativized semantics and thus of ontology.

3.24 Semantic Values

Semantic values are the elementary semantic component parts distributed among the meaning complexes associated with the descriptive terms of a language at a point in time.

For every descriptive term there are semantical rules with each rule's predicate describing component parts of the common subject term's meaning complex. A linguistic system therefore contains elementary components of meaning complexes that are shared by many descriptive terms, but are never uniquely associated with any single term, because all words have dictionary definitions analyzing the lexical entry's component parts. These elementary components may be called "**semantic values**".

Semantic values describe the most elementary ontological features of the real world that are distinguished by a language at a given point in time, and they are the smallest elements in any meaning complex at the given point in time. The indefinitely vast residual mind-independent reality not captured by any semantic values and that the language user's semantics is therefore unable to signify at the given point in time is due to the empirical underdetermination of the whole language at the time.

Different languages have different semantics and therefore display different ontologies. Where the semantics of one language displays semantic values not contained in the semantics of the other language, the two languages may be said to be semantically incommensurable. Translation is therefore made inexact, as has long been recognized by the refrain, "*traduttore, traditore*".

A science at different times in its history may also have semantically incommensurable language, when a later theory contains semantic values not contained in the earlier law or theory with even the same test design. But incommensurability does not occur in scientific revolutions understood as theory revisions, because the revision is a reorganization of pre-existing information. When incommensurability occurs, it occurs at times of discovery that occasion articulation of new semantic values due to new observations, even though the new observations may occasion a later theory revision.

3.25 Univocal and Equivocal Terms

A descriptive term's use is univocal, if no universally quantified negative categorical statement accepted as true can relate any of the predicates in the several universal affirmations functioning as semantical rules for the same subject term. Otherwise the term is equivocal.

If two semantical rules have the form "Every X is A" and "Every X is B", and if it is also believed that "No A is B", then the terms "A" and "B" symbolize parts of different meanings for the term "X", and "X" is equivocal. Otherwise

"A" and "B" symbolize different parts of the same meaning complex associated with the univocal term "X".

The definitions of descriptive terms such as common nouns and verbs in a unilingual dictionary function as semantical rules. Implicitly they are universally quantified logically, and are always presumed to be true. Usually each lexical entry in a large dictionary such as the *Oxford English Dictionary* offers several different meanings for a descriptive term, because terms are routinely equivocal. Language economizes on words by giving them several different meanings, which the fluent listener or reader can distinguish in context. Equivocations are the raw materials for puns (or for deconstructionist escapades). There is always at least one semantical rule for the meaning complex for each univocal use of a descriptive term, because to be meaningful, the term must be part of the linguistic system of beliefs. If the use is conventional, it must be capable of a lexical entry in a dictionary, or otherwise recognized by some trade or clique as part of their argot.

A definition, *i.e.,* a lexical entry in a unilingual dictionary functions as a semantical rule. But the dictionary definition is only a minimal description of the meaning complex of a univocal descriptive term, and it is seldom the whole description. Univocal terms routinely have many semantical rules, when many characteristics can be predicated in universally quantified beliefs to a given subject. Thus there are multiple predicates that universally characterize crows, characteristics known to the ornithologist, and which may fill a paragraph or more in his ornithological reference book.

Descriptive terms can become partially equivocal through time, when some parts of the term's meaning complex are unaffected by a change of defining beliefs, while other parts are simply dropped as archaic or are replaced by new parts contributed by new beliefs. In science this partial equivocation occurs when one theory is replaced by a newer one due to a test outcome, while the test designs for both theories remain the same. A term common to old and new theory may on occasion remain univocal only with respect to the parts contributed by the test-design language.

3.26 Signification and Supposition

Supposition enables identifying ambiguities not due to differences in signification that make equivocations, but instead are ambiguities due to differences in relating the semantics to its ontology.

The signification of a descriptive term is its meaning, and terms with two or more alternative significations are equivocal in the sense described above in Section **3.25.** The signification of a univocal term has different suppositions, when it describes, its ontology differently due to its having different functions in the sentences containing it.

Historically the subject term in the categorical proposition is said to be in "personal" supposition, because it references individual entities, while the predicate term is said to be in "simple" or "formal" supposition, because the predicate signifies attributes without referencing any individual entities manifesting the attributes. For this reason unlike the subject term the predicate term in the categorical proposition is not logically quantified with any syncategorematic quantifiers such as "every" or "some". For example in "Every crow is black" the subject term "crow" is in personal supposition, while the predicate "black" is in simple supposition; so too for "No crow is black".

The <u>subject-term</u> rôle in a sentence in object language has personal supposition, because it references entities.

The <u>predicate-term</u> rôle in a sentence in object language has simple or formal supposition, because it signifies attributes *<u>without</u>* referencing the entities manifesting the attributes.

Both personal and simple suppositions are types of "real" supposition, because they are different ways of talking about extramental nonlinguistic reality. They operate in expressions in object language and thus describe ontologies as either attributes or the referenced individuals characterized by the signified attributes.

In logical supposition the meaning of a term is considered specifically as a meaning.

Real supposition is contrasted with "logical" supposition, in which the meaning of the term is considered in the metalinguistic perspective exclusively as a meaning, *i.e.*, only semantics is considered and not extramental ontology. For example in "Black is a component part of the meaning of crow", the terms "crow" and "black" in this statement are in logical supposition. Similarly to say in explicit metalanguage "'Every crow is black' is a semantical rule" to express "Black is a component part of the meaning of crow", is again to use both "crow" and "black" in logical supposition.

Furthermore just to use "Every crow is black" as a semantical rule in order to exhibit its meaning composition without actually saying that it is a semantical rule, is also to use the sentence in the metalinguistic perspective and in logical supposition. The difference between real and logical supposition in such use of a sentence is not exhibited syntactically, but is pragmatic and depends on a greater context revealing the intention of the writer or speaker. Whenever a universally quantified affirmation is used in the metalinguistic perspective as a semantical rule for analysis in the semantical dimension, both the subject and predicate terms are in logical supposition. Lexical entries in dictionaries are in the metalinguistic perspective and in logical supposition, because they are about language and are intended to describe meanings.

In all the above types of supposition the same univocal term has the same signification. But another type of so-called supposition proposed incorrectly in ancient times is "material supposition", in which the term is referenced in metalanguage as a linguistic symbol in the syntactical dimension with no reference to a term's semantics or ontology. An example is "'Crow' is a four-letter word". In this example "crow" does not refer either to the individual real bird or to its characteristics as in real supposition or to the universal concept of the creature as in logical supposition. Thus material supposition is not supposition properly so called, because the signification is different. It is actually an alternative meaning and thus a type of semantical equivocation. Some philosophers have used other vocabularies for recognizing this equivocation, such as Stanislaw Leśniewski's (1886-1939) "use" (semantics) vs. "mention" (syntax) and Carnap's "material mode" (semantics) vs. "formal mode" (syntax).

3.27 Aside on Metaphor

A metaphor is a predication to a subject term that is intended to include only selected parts of the meaning complex conventionally associated with the predicate term, so the metaphorical predication is a true statement due to the exclusion of the remaining parts in the predicate's meaning complex that would conventionally make the metaphorical predication a false statement.

In the last-gasp days of decadent neopositivism some positivist philosophers invoked the idea of metaphor to explain the semantics of theoretical terms. And a few were closet Cartesians who used it in the charade of justifying realism for theoretical terms. The theoretical term was the positivists' favorite hobbyhorse. But both realism and the semantics of theories are unproblematic for contemporary pragmatists. In his "Posits and Reality" Quine said that all language is empirically underdetermined, and that the only difference between positing microphysical entities [like electrons] and macrophysical entities [like elephants] is that the statements describing the former are more empirically underdetermined than those describing the latter. Thus contrary to the neopositivists the pragmatists admit no qualitative dichotomy between the positivists' so-called observation terms and their so-called theoretical terms.

As science and technology advance, concepts of microphysical entities like electrons are made less empirically underdetermined, as occurred for example with the development of the cloud chamber. While contemporary pragmatist philosophers of science recognize no need to explain so-called theoretical terms by metaphor or otherwise, metaphor is nevertheless a linguistic phenomenon often involving semantical change, and it can easily be analyzed and explained with componential semantics.

It has been said that metaphors are both (unconventionally) true and (conventionally) false. In a speaker or writer's conventional or "literal" linguistic usage the entire conventional meaning complex associated with a univocal predicate term of a universal affirmation is operative. But in a speaker or writer's metaphorical linguistic usage only some selected component part or parts of the entire meaning complex associated with the univocal predicate term are operative, and the remaining parts of the meaning complex are intended to be excluded, *i.e.*, suspended from consideration and ignored. If the excluded parts were included, then the metaphorical statement would indeed be false. But the speaker or writer implicitly expects the hearer or reader to recognize and suspend from consideration the excluded parts of the predicate's conventional semantics, while the speaker or writer uses the component part that he has tacitly selected for describing the subject truly.

Consider for example the metaphorical statement "Every man is a wolf." The selected meaning component associated with "wolf" that is intended to be predicated truly of "man" might describe the wolf's predatory behaviors, while the animal's quadrupedal anatomy, which is conventionally associated with "wolf", is among the excluded meaning components for "wolf" that are not intended to be predicated truly of "man".

A listener or reader may or may not succeed in understanding the metaphorical predication depending on his ability to select the applicable parts of the predicate's semantics tacitly intended by the issuer of the metaphor. But there is nothing arcane or mysterious about metaphors, because they can be explained in "literal" (*i.e.*, conventional) terms to the uncomprehending listener or reader. To explain the metaphorical predication of a descriptive term to a subject term is to list explicitly those affirmations intended to be true of that subject and that set forth just those parts of the predicate's meaning that the issuer of the metaphor intends to be applicable.

The explanation may be further elaborated by listing separately the affirmations that are not viewed as true of the subject, but which are associated with the predicated term when it is predicated conventionally. Or these may be expressed as universal negations stating what is intended to be excluded from the predicate's meaning complex in the particular metaphorical predication, *e.g.*, "No man is quadrupedal." In fact such negative statements might be given as hints by a picaresque issuer of the metaphor for the uncomprehending listener.

A semantical change occurs when the metaphorical predication becomes conventional, and this change to conventionality produces an equivocation. The equivocation consists of two "literal" meanings: the original one and a derivative meaning that is now a dead metaphor. As a dead man is no longer a man, so a dead metaphor is no longer a metaphor. A dead metaphor is a meaning from which the suspended parts in the metaphor have become conventionally excluded to produce a new "literal" meaning. Trite metaphors, when not just forgotten, metamorphose into new literals, as they eventually become conventional.

There is an alternative "interactionist" concept of metaphor that was proposed by Max Black (1909-1988), a Cambrian positivist philosopher, in his *Models and Metaphors* (1962). On Black's interactionist view both the subject and predicate terms change their meanings in the metaphorical statement due to a semantical "interaction" between them. Black does not describe the process of interaction. Curiously he claims for example that the metaphorical statement "Man is a wolf" allegedly makes wolves seem more human and men seem more lupine. This is merely obscurantism; it is not logical, because the statement "Every man is a wolf" in not universally convertible; "Every man is a wolf" does not imply logically "Every wolf is a man". The metaphorical use of "wolf" in "Every man is a wolf" therefore does not make the subject term "man" a metaphor. "Man" becomes a metaphor only if there is an independent acceptance of "Every wolf is a man", where "man" occurs as a predicate.

3.28 Clear and Vague Meaning

Vagueness is empirical underdetermination, and can never be eliminated completely, since our concepts can never grasp reality exhaustively.

Meanings are more or less clear and vague, such that the greater the clarity, the less the vagueness. In "Verifiability" in *Logic and Language* (1952) Friedrich Waismann (1896-1954) called this inexhaustible residual vagueness the "open texture" of concepts.

Vagueness in the semantics of a univocal descriptive term is reduced and clarity is increased by the addition of universal affirmations and/or negations accepted as true, to the list of the term's semantic rules with each rule having the term as a common subject. The clarification is supplied by the semantics of the predicates in the added universal affirmations and/or negations.

Additional semantical rules increase clarity. If the list of universal statements believed to be true are in the form "Every X is A" and "Every X is B", then clarification of X with respect to a descriptive predicate "C" consists in adding to the list either the statement in the form "Every X is C" or the statement in the form "No X is C". Clarity is thereby added by amending the meaning of "X".

Clarity is also increased by adding semantical rules that relate any of the univocal predicates in the list of semantical rules for the same subject thus increasing coherence.

If the predicate terms "A" and "B" in the semantical rules with the form "Every X is A" and "Every X is B" are related by the statements in the form "Every A is B" or "Every

B is A", then one of the statements in the expanded list can be logically derived from the others by a syllogism. Awareness of the deductive relationship and the consequent display of structure in the meaning complex associated with the term "X" makes the complex meaning of "X" more coherent, because the deductive relation makes it more semantically integrated thus enhancing coherence. Clarity is thereby added by exhibiting semantic structure in a deductive system. And the resulting coherence also supplies a psychological satisfaction, because people prefer to live in a coherent world. However "Every A is B" is also an empirical statement that may be falsified, and if it is tested and not falsified, it offers more than psychological satisfaction, because it is a new law.

These additional semantical rules relating the predicates may be negative as well as affirmative. Additional universal negations offer clarification by exhibiting equivocation. Thus if two semantical rules are in the form "Every X is A" and "Every X is B", and if it is also believed that "No A is B" or its equivalent "No B is A", then the terms "A" and "B" symbolize parts of different meanings for the term "X", and "X" is equivocal. Clarity is thereby added by the negation.

3.29 Semantics of Mathematical Language

The semantics for a descriptive mathematical variable intended to take measurement values is determined by its context consisting of universally quantified statements believed to be true including mathematical expressions in the theory language proposed for testing and in the test-design language presumed for testing.

Both test designs and theories often involve mathematical expressions. Thus the semantics for the descriptive variables common to a test design and a theory may be supplied wholly or in part by mathematical expressions, such that the structure of their meaning complexes is partly mathematical. The semantics-determining statements in test designs for mathematically expressed theories may include mathematical equations, measurement language describing the subject measured, the measurement procedures, the metric units and any employed apparatus.

Some of these statements may resemble what 1946 Nobel-laureate physicist Percy Bridgman (1882-1961) in his *Logic of Modern Physics* (1927) calls "operational definitions", because the statements describing the measurement procedures and apparatus contribute meaning to the descriptive term that occurs in a test design. Bridgman says that a concept is synonymous with a corresponding set of operations. But contrary to Bridgman and as even the positivist Carnap recognized in his *Philosophical Foundations of Physics* (1966), each of several operational definitions for the same term does not constitute a separate definition for the term's concept of the measured subject, thereby making the term equivocal. Likewise pragmatists say that descriptions of different measurement procedures contribute different parts to the meaning of the univocal descriptive term, unless the different procedures produce different measurement values, where the differences are greater than the estimated measurement errors in the overlapping ranges of measurement. Furthermore contrary to Bridgman operational definitions have no special status; they are just one of many possible types of statement often found in a test design.

3.30 Semantical State Descriptions

A semantical state description for a scientific profession is a synchronic display of the semantical composition of the various meanings of the partially equivocal descriptive terms in the several alternative

theories functioning as semantical rules and addressing a single problem defined by a common test design.

The above discussions in philosophy of language have focused on descriptive terms such as words and mathematical variables, and then on statements and equations that are constructed with the terms. For computational philosophy of science there is an even larger unit of language, which is the semantical state description.

In his *Meaning and Necessity* (1947) Carnap had introduced a concept of semantical state description in his philosophy of semantical systems. Similarly in computational philosophy of science a state description is a semantical description but different from Carnap's. Both the statements and/or equations supplying the terms for a discovery system's input state description and the statements constituting the output state description are all semantical rules. Each alternative theory or law in the state description has its distinctive semantics for its constituent descriptive terms. A term shared by several alternative theories or laws is thus partly equivocal. But the term is also partly univocal due to the common test-design statements that are also semantical rules, which are operative in both input and output state descriptions.

In computational philosophy of science the state description is a synchronic and thus a static semantical display. The state description contains language actually used in a science both in an **initial state description** supplying object-language terms inputted to a discovery system, and in a **terminal state description** containing new object-language statements output generated by a computerized discovery-system's execution. The initial state description represents the current frontier of research for the specific problem. Both input and output state descriptions for a discovery-system execution address only one problem identified by the common test design, and thus for computational philosophers of science they represent only one scientific "profession" (See below, Section **3.47**).

A discovery-system is a mechanized finite-state generative grammar that produces sentences or equations from inputted descriptive terms or variables. As a grammar it is creative in Noam Chomsky's (1928) sense in his *Syntactical Structures* (1957), because when encoded in a computer language and executed, the system produces new statements, i.e. theories that have never previously been stated in the particular scientific profession. A mechanized discovery system is Feyerabend's principle of theory proliferation applied with mindless abandon. But to control the size and quality of the output, the system also tests the empirical adequacy of the generated novel theories and inevitably rejects most of them. Associating measurement data with the inputted variables enables empirical testing, so that the system designs often employ one or another type of applied numerical methods.

For semantical analysis a state description consists of universally quantified statements and/or equations. The statements and/or equations including theories and the test design from which the inputted terms were extracted are included in the state description although not for discovery system input, because they would prejudice the output. Statements and/or equations function as semantical rules in the generated output only. Thus for discovery-system input, the input is a listing of descriptive terms extracted from the statements and/or equations of the several currently untested theories addressing the same unsolved problem as defined by a common test design at a given point in time.

Descriptive terms extracted from the statements and/or equations constituting falsified theories can also be included to produce a cumulative state description for input, because the terms from previously falsified theories represent available information at the historical or current point in time. Descriptive terms salvaged from falsified theories have scrap value, because they may be recycled productively through the theory-developmental process. Furthermore terms and variables from tested and nonfalsified theories could also conceivably be included, just to see what new comes out. Empirical underdetermination permits scientific pluralism, and the world is full of surprises.

3.31 Diachronic Comparative-Static Analysis

A diachronic comparative-static display consists of two chronologically successive state descriptions containing theory statements for the same problem defined by the same test design and therefore addressed by the same scientific profession.

State descriptions contain statements and equations that operate as semantical rules displaying the meanings of the constituent descriptive terms and variables. Comparison of the statements and equations in two chronologically separated state descriptions containing the same test design for the same profession exhibits semantical changes resulting from the transition.

In computational philosophy of science this comparison is typically a comparison of a discovery system's originating input and generated output state descriptions of theory statements for purposes of contrast.

3.32 Diachronic Dynamic Analysis

The dynamic diachronic metalinguistic analysis not only consists of two state descriptions representing two chronologically successive language states sharing a

common subset of descriptive terms in their common test design, but also describes a process of linguistic change between the two successive state descriptions.

Such transitions in science are the result of two pragmatic functions in basic research, namely **theory development** and **theory testing.** A change of state description into a new one is produced whenever a new theory is proposed or whenever a theory is eliminated by a falsifying test outcome.

3.33 Computational Philosophy of Science

Computational philosophy of science is the development of mechanized discovery systems that can explicitly proceduralize and thus mechanize a transition applied to the current state description of a science, in order to develop a new state description containing one or several new and empirically adequate theories.

The discovery systems created by the computational philosopher of science represent diachronic dynamic metalinguistic analyses. The systems proceduralize developmental transitions explicitly with a mechanized system design, in order to accelerate the advancement of a contemporary state of a science. Their various procedural system designs are metalinguistic logics for rational reconstructions of scientific discovery. By applying the system to the vocabulary in the current state description for the science the systems generate new theories. The discovery systems typically include empirical criteria for selecting a subset of the generated theories for output as tested and nonfalsified theories either for further predictive testing or for use as laws in explanations and test designs.

But presently few philosophy professors have the needed competencies to contribute to computational philosophy of science, because few curricula in university philosophy departments even encourage much less actually prepare students for contributing to this new and emerging area in philosophy of science for their necessarily interdisciplinary Ph.D. dissertations. Among today's academic philosophers the mediocrities will simply ignore this new area, while the Luddites will shrilly reject it. Lethargic and/or reactionary academics that dismiss it are fated to spend their careers denying its merits and evading it, as they are inevitably marginalized, destined to die in obscurity. But the exponentially growing capacities of computer hardware and the proliferation of computer-systems designs have already been enhancing the practices of basic-scientific research in many sciences.

Thus in his *Extending Ourselves* (2004) University of Virginia philosopher of science and cognitive scientist Paul Humphreys reports that computational science for scientific analysis has already far outstripped natural human capabilities and that it currently plays a central rôle in the development of many physical and life sciences. Neither philosophy of science nor the retarded social sciences can escape such developments. Computational philosophy of science is already achieving ascendancy in twenty-first-century philosophy of science due to those who are opportunistic enough to master both the necessary computer skills and the requisite working competencies in an empirical science.

In the "Introduction" to their *Empirical Model Discovery and Theory Evaluation*: *Automatic Selection Methods in Econometrics* (2014) David F. Hendry and Jurgen A. Doornik of Oxford University's "Program for Economic Modeling at their Institute for New Economic Thinking" write that automatic modeling has "come of age." Hendry was head of Oxford's Economics Department from 2001 to 2007, and is presently Director of the Economic Modeling Program at

Oxford University's Martin School. These authors have developed a mechanized general-search algorithm that they call **AUTOMETRICS** for determining the equation specifications for econometric models.

Computational philosophy of science is the future that has arrived, even when it is called by other names as practiced by scientists working in their special fields instead of being called "metascience", "computational philosophy of science" or "artificial intelligence". Our twenty-first century perspective shows that computational philosophy of science has indeed "come of age", as Hendry and Doornik report.

Artificial intelligence today is producing an institutional change in the sciences and humanities. In "MIT Creates a College for Artificial Intelligence, Backed by $1 Billion" *The New York Times* (16 October 2018) reported that the Massachusetts Institute of Technology will create a new college with fifty new faculty positions and many fellowships for graduate students, in order to integrate artificial intelligence systems into both its humanities and its science curricula. The article quoted L. Rafael Reif, president of MIT as stating that he wanted artificial intelligence to make a university-wide impact and to be used by everyone in every discipline [presumably including philosophy of science]. And the article also quoted Melissa Nobles, dean of MIT's School of Humanities and Sciences, as stating that the new college will enable the humanities to survive, not by running from the future, but by embracing it. So, there is hope that the next generation of journal editors with their favorite referees, whose peer-reviewed publications now operate as a refuge for Luddites, reactionaries and hacks, will stop running from the future and belatedly acknowledge the power and productivity of artificial intelligence.

3.34 An Interpretation Issue

There is ambiguity in the literature as to what a state description represents and how the discovery system's processes are to be interpreted. The phrase "artificial intelligence" has been used in both interpretations but with slightly different meanings.

On the linguistic analysis interpretation, which is the view taken herein the state description represents the language state for a language community constituting a single scientific profession identified by a test design. Like the diverse members of a profession, the system produces a diversity of new theories. But no psychological claims are made about intuitive thinking processes.

Computer discovery systems are generative grammars that generate and test theories.

The computer discovery systems are mechanized generative grammars that construct and test theories. The system inputs and outputs are both object-language state descriptions. The instructional code of the computer system is in the metalinguistic perspective, and exhibits diachronic dynamic procedures for theory development. The various procedural discovery system designs are the logics for rational reconstructions of discovery. As such the linguistic analysis interpretation is neither a separate philosophy of science nor a psychologistic agenda. It is compatible with the contemporary pragmatism and its use of generative grammars makes it closely related to computational linguistics.

On the cognitive-psychology interpretation the state description represents a scientist's cognitive state consisting of mental representations and the discovery system represents the scientist's cognitive processes.

Computer discovery systems are psychological hypotheses about intuitive human problem-solving processes.

The originator of the cognitive-psychology interpretation is Simon. In his *Scientific Discovery: Computational Explorations of the Creative Processes* (1987) and other works Simon writes that he seeks to investigate the psychology of discovery processes, and to provide an empirically tested theory of the information-processing mechanisms that are implicated in that process. There he states that an empirical test of the systems as psychological theories of human discovery processes would involve presenting the computer programs and some human subjects with identical problems, and then comparing their behaviors. But Simon admits that his book provides nothing by way of comparison with human performance. And in discussions of particular applications involving particular historic discoveries, he also admits that in some cases the historical scientists actually performed their discoveries differently than the way the systems performed the rediscoveries.

The academic philosopher Paul Thagard, who also follows Simon's interpretation, originated the name "computational philosophy of science" in his *Computational Philosophy of Science* (1988). Hickey admits that it is more descriptive than the name "metascience" that Hickey had proposed in his *Introduction to Metascience* a decade earlier. Thagard defines computational philosophy of science as "normative cognitive psychology". His cognitive-psychology systems have successfully replicated developmental episodes in history of science, but the relation of their system designs to systematically observed human cognitive processes is still unexamined. And their outputted theories to date have not yet proposed any new contributions to the current state of any science.

In “A Split in Thinking among Keepers of Artificial Intelligence”(AI) *The New York Times* (18 July 1993) reported that scientists attending the annual meeting of the American Association of Artificial Intelligence expressed disagreement about the goals of artificial intelligence. Some maintained the traditional view that artificial-intelligence systems should be designed to simulate intuitive human intelligence, while others maintained that the phrase “artificial intelligence” is merely a metaphor that has become an impediment, and that AI systems should be designed to exceed the limitations of intuitive human intelligence.

So, is artificial intelligence computerized psychology or computerized linguistics? There is no unanimity. To date the phrase “computational philosophy of science” need not commit one to either interpretation. Which interpretation prevails in academia will likely depend on which academic department productively takes up the movement. If the psychologists develop new and useful systems, the psychologistic interpretation will prevail. If the philosophers take it up successfully, their linguistic-analysis interpretation will prevail.

In their “Processes and Constraints in Explanatory Scientific Discovery” in *Proceedings of the Thirteenth Annual Meeting of the Cognitive Science Society* (2008) Langley and Bridewell, who advocate Simon’s cognitive-psychology interpretation, appear to depart from the cognitive-psychology interpretation. They state that they have not aimed to “mimic” the detailed behavior of human researchers, but that instead their systems address the same tasks as scientists and carry out search through similar problem spaces. This much might also be said of the linguistic-analysis approach.

For more about Simon, Langley, and Thagard and about discovery systems and computational philosophy of science readers are referred to **BOOK VIII** at the free web site www.philsci.com or in the e-book *Twentieth-Century Philosophy of Science: A History,* which is available from most Internet booksellers.

C. ONTOLOGY

3.35 Ontological Dimension

Ontology is the aspects of mind-independent reality revealed by relativized perspectivist semantics.

Semantics is description of reality; ontology is reality as described by semantics.

Ontology is the metalinguistic dimension after syntax and semantics, and it presumes both of them. It is the reality that is signified by semantics. Semantically interpreted syntax describes ontology most realistically, when the statement is warranted empirically by independently repeated nonfalsifying test outcomes. Thus in science ontology is more adequately realistic, when described by the semantics of either a scientific law or an observation report having its semantics defined by a law. The semantics of falsified theories display ontology less realistically due to the falsified theories' demonstrated lesser empirical adequacy.

3.36 Metaphysical and Scientific Realism

Metaphysical realism is the thesis that there exists mind-independent reality, which is accessible to and accessed by human cognition.

Few scientists will deny that their explanations describe reality; indeed the belief is integral to their motivation to practice basic research. Yet traditional philosophers have spilt much wasted ink arguing over realism and its alternatives. The above statement of metaphysical realism is disarmingly simple. It is simply the affirmation of reality and its accessibility with human knowledge. Most importantly "metaphysical realism" does not mean any characterization of reality like some super ontology that can be described by any transcendental, all-encompassing or (to use Hilary Putnam's phrase) "God's-Eye View". It is recognition that there exists mind-independent reality responsible for both recalcitrantly falsifying empirical tests and everyday surprises.

In the section titled "Is There Any Justification for External Realism" in his *Mind, Language and Society: Philosophy in the Real World* (1995) University of California realist philosopher John R. Searle (1932) refers to metaphysical realism as "external realism", by which he means that the world exists independently of our representations of it. He says that realism does not say how things are, but only that there is a way that they are. The way that they are would include Heisenberg's "*potentia*" as the quantum theory describes reality with its indeterminacy relations. The theory describes microphysical reality as being that certain way and not otherwise, such that the theory is testable and falsifiable although not yet falsified.

Searle denies that external realism can be justified, because any attempt at justification presupposes what it attempts to justify. In other words all arguments for metaphysical realism are circular, because realism must firstly be accepted. Any attempt to find out about the real world presupposes that there is a way that things are. He goes on to affirm the picture of science as giving us knowledge of independently existing reality, and that this picture is taken for granted in the sciences.

Similarly in “Scope and Language of Science” in *Ways of Paradox* (1976) Harvard University realist philosopher Quine writes that we cannot significantly question the reality of the external world or deny that there is evidence of external objects in the testimony of our senses, because to do so is to dissociate the terms “reality” and “evidence” from the very application that originally did most to invest these terms with whatever intelligibility they may have for us. And to emphasize the primal origin of realism Quine writes that we imbibe this primordial awareness “with our mother’s milk”. He thus affirms what he calls his “unregenerate realism”. These statements by Searle, Quine and others of their ilk are not logical arguments or inferences; they are simply affirmations.

And Heidegger too recognized that the problem of the reality of the external world is a pseudo problem. He notes that while for Kant the scandal of philosophy is that no proof has yet been given of the existence of things outside of us, for Heidegger the scandal is not that this proof has yet to be given, but that any such proof is expected and that it is often attempted.

Hickey joins these contemporary realist philosophers. He maintains that metaphysical realism, the thesis that there exists self-evident mind-independent reality accessible to and accessed by cognition, is the “primal prejudice” that cannot be proved or disproved but can only be affirmed or denied. Mind-independent reality is not Kant’s ineffable reality, but rather is the very effable reality revealed by our perspectivist semantics as ontologies. And he affirms that the primal prejudice is a correct and universal prejudice, even though there are delusional psychotics and sophistic academics that are in denial.

Contrary to Descartes and latter-day rationalists, metaphysical realism is neither a conclusion nor an inference nor an extrapolation. It cannot be proved logically, established by philosophy or science, validated or justified in any discursive manner including figures of speech such as analogy or metaphor. Its self-evident character makes it antecedent to any such discursive judgments by the mind, and therefore fundamentally prejudicial. Hickey regards misguided pedantics who say otherwise as "closet Cartesians", because they never admit they are neo-Cartesians. The imposing, intruding, recalcitrant, obdurate otherness of mind-independent reality is immediately self-evident at the dawn of a person's consciousness; it is the most rudimentary experience. Dogs and cats, bats and rats, and all the other sentient creatures that have survived Darwinian predatory reality are infra-articulate and nonreflective realists in their apprehensions of their environments. To dispute realism is to step through the looking glass into Alice's labyrinth of logomanchy, of metaphysical jabberwocky where as Schopenhauer believed the world is but a dream. It is to indulge in the philosophers' hallucinatory escapist narcotic.

Scientific realism is the thesis that a tested and currently nonfalsified theory describes the most empirically adequate, the truest and thus most realistic ontology at the current time.

After stating that the notion of reality independent of language is in our earliest impressions, Quine adds that it is then carried over into science as a matter of course. He writes that realism is the robust state of mind of the scientist, who has never felt any qualms beyond the negotiable uncertainties internal to his science.

N.B. Contrary to Feyerabend the phrase "scientific realism" does not mean scientism, the thesis that only science describes reality.

3.37 Ontological Relativity Defined

When metaphysical realism is joined with relativized semantics, the result is ontological relativity.

Ontological relativity in science is the thesis that the perspectivist semantics of a theory or law and its constituent descriptive terms describe aspects of mind-independent reality.

The ontology of any theory or law is as realistic as it is empirically adequate.

Understanding scientific realism requires consideration of ontological relativity. Ontological relativity is the subordination of ontology to empiricism. We cannot separate ontology from semantics, because we cannot step outside of our knowledge and compare our knowledge with reality, in order to validate a correspondence. But we can distinguish our semantics from the ontology it reveals, as we do for example, when we distinguish logical and real suppositions respectively in statements. We describe mind-independent reality with our perspectivist semantics, and ontology is reality as it is thus revealed empirically more or less adequately by our semantics. Our semantics and thus ontologies cannot be exhaustive, but ontologies are more or less adequately realistic, as the semantics is more or less adequately empirical.

Prior to the evolution of contemporary pragmatism philosophers had identified realism as such with one or another particular ontology, which they erroneously viewed as the only ontology on the assumption that there can be only one ontology. Such is the error made by some physicists who believe that they are defending realism, when they defend the "hidden variable" interpretation of quantum theory. Such too is the error in Popper's proposal for his propensity interpretation of quantum theory. Similarly contrary to Einstein's EPR thesis of a single uniform ontology for physics, Aspect, Dalibard and Roger's findings from their 1982 nonlocality experiments demonstrated entanglement empirically and thus validated the Copenhagen interpretation's semantics and ontology.

Advancing science has produced revolutionary changes. And as the advancement of science has produced new theories with new semantics exhibiting new ontologies, some prepragmatist scientists and philosophers found themselves attacking a new theory and defending an old theory, because they had identified realism with the ontology associated with the older falsified theory. As Feyerabend notes in his *Against Method*, scientists have criticized a new theory using the semantics and ontology of an earlier theory. Such a perversion of scientific criticism is still common in the social sciences where romantic ontologies are invoked as criteria for criticism.

With ontological relativity realism is no longer uniquely associated with any one particular ontology. The ontological-relativity thesis does not deny metaphysical realism, but depends on it. It distinguishes the mind-independent plenitude from the ontologies revealed by the perspectivist semantics of more or less empirically adequate beliefs. Ontological relativity enables admitting change of ontology without resorting to instrumentalism, idealism, phenomenalism, solipsism, any of the several varieties of antirealism, or any other such denial of metaphysical realism.

Thus ontological relativity solves the modern problem of reconciling conceptual revision in science with metaphysical realism. Ontological relativity enables acknowledging the creative variability of knowledge operative in the relativized semantics and consequently mind-dependent ontologies that are defined in constructed theories, while at the same time acknowledging the regulative discipline of mind-independent reality operative in the empirical constraint in tests with their possibly falsifying outcomes.

In summary, in contemporary pragmatist philosophy of science metaphysical realism is logically prior to and presumed by all ontologies as the primal prejudice, while the choice of an ontology is based upon the empirically demonstrated adequacy of the theory describing the ontology. Indulging in futile disputations about metaphysical realism will not enhance achievement of the aims of either science or philosophy of science, nor will dismissing such disputations encumber achieving those aims. Ontological relativity leaves ontological decisions to the scientist rather than the metaphysician. And the superior empirical adequacy of a new law yields the increased truth of a new law and the increased realism in the ontology that the new law reveals.

3.38 Ontological Relativity Illustrated

There is no semantically interpreted syntax that does not reveal some more or less realistic ontology.

Since all semantics is relativized, is part of the linguistic system, and ultimately comes from sense stimuli, no semantically interpreted syntax – not even the description of a hallucination – is utterly devoid of ontological significance.

To illustrate ontological relativity consider the semantical decision about white crows mentioned in the above discussion about componential artifactual semantics (See above, Section **3.20** and **3.23**). The decision is ontological as well as semantical. For the bird watcher who found a white but otherwise crow-looking bird and decides to reject the belief "Every crow is black", the phrase "white crow" becomes a description for a type of existing birds. Once that semantical decision is made, white crows suddenly populate many trees in the world, however long ago Darwinian Mother Nature had evolved the observed avian creatures. But if his decision is to persist in believing "Every crow is black", then there are no white crows in existence, because whatever kind of creature the bird watcher found and that Darwinian Mother Nature had long ago evolved, the white bird is not a crow. The availability of the choice illustrates the artifactuality of the relativized semantics of language and of the consequently relativized ontology that the relativized semantics reveals about mind-independent reality.

Relativized semantics makes ontology no less relative whether the affirmed entity is an elephant, an electron, or an elf. Beliefs that enable us routinely to make successful predictions are deemed more empirically adequate and thus more realistic and truer than those less successfully predictive. And we recognize the reality of the entities, attributes or any other characteristics that enable those routinely successful predicting beliefs. Thus if positing evil elves conspiring mischievously enabled predicting the collapse of market-price bubbles on Wall Street more accurately and reliably than the postulate of euphoric humans speculating greedily, then we would decide that the ontology of evil elves is as adequately realistic as it was found to be adequately empirical, and we would busy ourselves investigating elves, as we would do with elephants and electrons for successful predictions about elephants and electrons. On the other hand were our price predictions to fail, then those failures would inform us that our belief in the elves of Wall Street is as empirically inadequate as the discredited belief in the legendary gnomes of Zürich that are reputed to manipulate currency speculations ruinous to the wealth of nations, and we would decide that the ontology of elves is as inadequately realistic, as it was inadequately empirical.

Consider another illustration. Today we reject an ontology of illnesses due to possessing demons as inadequately realistic, because we do not find ontological claims about possessing demons to be empirically adequate for effective medical practice. But it could have been like the semantics of "atom". The semantics and ontology of "atom" have changed greatly since the days of the ancient philosophers Leucippus and Democritus. The semantics of "atom" has since been revised repeatedly under the regulation of empirical research in physics, as when 1906 Nobel laureate J.J. Thomson discovered that the atom has internal structure, and thus today we still accept a semantics and ontology of atoms.

Similarly the semantics of "demon" might too have been revised to become as beneficial as the modern meaning of "bacterium", had empirical testing regulated an evolving semantics and ontology of "demon". Both ancient and modern physicians may observe and describe some of the same symptoms for a certain disease in a sick patient and both demons and bacteria are viewed as living agents, thus giving some continuity to the semantics and ontology of "demon" through the ages. But today's physicians' medical understanding, diagnoses and remedies are quite different. If the semantics and ontology of "demon" had been revised under the regulation of increasing empirical adequacy, then today scientists might materialize (*i.e.*, visualize) demons with microscopes, physicians might write incantations (*i.e.*, prescriptions), and pharmacists might dispense antidemonics (*i.e.*, antibiotics) to exorcise (*i.e.*, to cure) possessed (*i.e.*, infected) sick persons. But then terms such as "materialize", "incantation", "antidemonics", "exorcise" and "possessed" would also have acquired new semantics in the more empirically adequate modern contexts than those of ancient medical beliefs. And the descriptive semantics and ontology of "demon" would have been revised to exclude what we now find empirically to be inadequately realistic, such as a demon's willful malevolence.

This thesis can be found in Quine's "Two Dogmas of Empiricism" (1952) in his *Logical Point of View* (1953) even before he came to call it "ontological relativity" sixteen years later. There he says that physical objects are conceptually imported into the linguistic system as convenient intermediaries, as irreducible posits comparable epistemologically to the gods of Homer. But physical objects are epistemologically superior to other posits including the gods of Homer, because the former have proved to be more efficacious as a device for working a manageable structure into the flux of experience. And as a realist, he might have added explicitly that experience is experience **of** something, and that physical objects are more efficacious than whimsical gods for making correct predictions.

Or consider the tooth-fairy ontology. In some cultures young children losing their first set of teeth are told that if they place a lost tooth under the pillow at bedtime, an invisible tooth-fairy person having large butterfly wings will exchange the tooth for a coin as they sleep. The boy who does so and routinely finds a coin the next morning, has an empirically warranted belief in the semantics describing an invisible winged person that leaves coins under pillows and is called a "tooth fairy". This belief is no less empirical than belief in the semantics positing an invisible force (gravitons?) that pulls apples from their trees to the ground and is called "gravity". But should the child forget to advise his mother that he placed a recently lost tooth under his pillow, he will rise the next morning to find no coin. The boy's situation is complicated, because the concept of tooth fairy is not simply unrealistic; no semantically interpreted syntax is utterly devoid of ontological significance. In this case the boy has previously seen insects with butterfly wings, and there was definitely someone who swapped coins for the boy's lost teeth on previous nights. Yet the tooth fairy's recent nondelivery of a coin has given him reason to be suspicious about the degree of realism in the tooth fairy idea.

Thus like the bird watcher with a white crow-looking bird, the boy has semantical and ontological choices. He may continue to define "tooth fairy" as a benefactor other than his mother, and reject the tooth-fairy semantics and ontology as inadequately realistic. Or like the ancient astronomers who concluded that the morning star and the evening star are the same luminary and not stellar (*i.e.*, the planet Venus), he may revise his semantics of "tooth fairy" to conclude that his mother and the tooth fairy are the same benefactor and not winged. But later when he publicly calls his mother "tooth fairy", he will likely be encouraged to revise this semantics of "tooth fairy" again, and to accept the more conventional semantics and ontology that excludes tooth fairies, as modern physicians exclude willfully malevolent demons. This sociology of knowledge and ontology has been insightfully examined by the sociologists of knowledge Peter Berger (1929-2017) and Thomas Luckmann (1927-2016) in *The Social Construction of Reality* (1966).

Or consider ontological relativity in fictional literature. "Fictional ontology" is an oxymoron. But fictional literature resembles metaphor, because its discourse is recognized as having both true and false aspects (See above, Section **3.27**). For fictional literature the reader views as true the parts of the text that reveal reality adequately, and the reader excludes as untrue the parts that he views critically and finds to be inadequately realistic.

Readers know that Huckelbery Finn is a fictitious creation of Samuel Clemens (1835-1910), a.k.a. Mark Twain, but they also know that white teenagers with Huck's racist views existed in early nineteenth-century antebellum Southern United States. Sympathetic readers, who believe Twain's portrayal of the slavery ontology, recognize an ontology that is realistic about the injustices of the racist antebellum South. And initially unsympathetic readers who upon reading Twain's portrayal of Huck's dawning awareness of fugitive slave Jim's humanity notwithstanding Huck's racist upbringing, may thus be led to accept the more realistic ontology of black persons that is without the dehumanizing fallacies of racism. Ontological relativity enables recognition that such reconceptualization can reveal a more realistic ontology not only in science but in all discourse including even fiction.

Getting back to science, consider the Eddington eclipse test of Einstein's relativity theory mentioned above in the discussion of componential semantics (See above, Section **3.22**). That historic astronomical test is often said to have "falsified" Newton's theory. Yet today the engineers of the U.S. National Aeronautics and Space Administration (NASA) routinely use Newton's physics to navigate interplanetary rocket flights through our solar system. Thus it must be said that Newton's "falsified" theory is not completely false or unrealistic, or neither NASA nor anyone else could ever have used it. Therefore the Newtonian ontology must be realistic, but it is now known to be less realistic and more empirically underdetermined than the Einsteinian ontology, because the former has been demonstrated to be less empirically adequate.

3.39 Causality

Cause and effect are ontological categories, which in science can be described by tested and nonfalsified nontruth-functional hypothetical-conditional statements thus having the status of laws. The nontruth-functional hypothetical-conditional law statement claiming a causal dependency is an empirical universally quantified statement. It is therefore never proved and is always vulnerable to future falsification. But ontological relativity means that a statement's empirical adequacy warrants belief in its ontological claim of causality, even when the relation is stochastic. Nonfalsification does not merely make the statement affirm a Humean constant psychological conjunction. When in the progress of science an empirically tested causal claim is not empirically falsified, it is made evident thereby that the causality claim is more adequately true and thus more realistic than previously hypothesized.

Tested and nonfalsified correlation indicates causality, until the correlation is later empirically invalidated experimentally or otherwise experientially.

3.40 Ontology of Mathematical Language

In the categorical proposition the logically quantified subject term references individuals and describes the attributes that enable identifying the referenced individuals, while the predicate term describes only attributes without referencing the instantiated individuals manifesting the attributes. The referenced real entities and their semantically signified real attributes constitute the ontology described by the categorical proposition that is believed to be true due to its experimentally or otherwise experientially demonstrated empirical adequacy. These existential conditions are expressed explicitly by the copula term "is" as in "Every crow **is** black".

However, the ontological claim made by the mathematical equation in science is not only about instantiated individuals or their attributes. The individual instances referenced by the descriptive variables in the empirical mathematical equation are also instances of individual measurement results, which are magnitudes determined by comparison with some standard that are acquired by executing measurement procedures yielding numeric values for the descriptive variables. The individual measurement results are related to the measured reality by nonmathematical language, which includes description of the measured subject, the chosen metric, the measurement procedures, and any employed apparatus, all of which are included in a test design.

Also calculated and predicted values for descriptive variables describing effects in equations with measurement values for other variables describing causal factors, make ontological claims that are tested empirically. Untested theories make relatively more hypothetical quantitative causal claims. Tested and nonfalsified empirical equations are quantitative causal laws, unless and until they are eventually falsified.

D. PRAGMATICS

3.41 Pragmatic Dimension

Pragmatics is the uses or functions of language. The pragmatics of basic research in science is theory construction and empirical testing, in order to produce laws for explanations and test designs.

Pragmatics is the metalinguistic dimension after syntax, semantics and ontology, and it presupposes all of them. The regulating pragmatics of basic science is set forth in the statement of the aim of science, namely to create explanations containing scientific laws by development and empirical testing of theories, which are deemed laws when not falsified by the currently most critically empirical test. Explanations and laws are accomplished science, while theories and tests are work in progress at the frontier of basic research. Understanding the pragmatics of science therefore requires understanding theory development and testing.

3.42 Semantic Definitions of Theory Language

For the extinct neopositivist philosophers the term "theory" referred to universally quantified sentences containing "theoretical terms" that reference unobserved phenomena or entities.

The nineteenth-century positivists such as the physicist Ernst Mach rejected theory, especially the atomic theory of matter in physics, because atoms were deemed unobservable. These early positivist philosophers' idea of discovery consisted of induction, which yields empirical generalizations rather than theories containing theoretical terms.

Later the twentieth-century neopositivists believed that they could validate the meaningfulness of theoretical terms referencing unobserved microphysical particles such as electrons, and thus admit theories as valid science. For discovery of theories they invoked human creativity but offered no description of the processes of theory creation.

These neopositivists also viewed Newton's physics as paradigmatic of theoretical science. They therefore also construed "theory" to mean an axiomatic system, because Kepler's laws of orbital motion can be derived deductively as theorems from Newton's inverse-square principle.

For the anachronistic romantic philosophers and romantic social scientists "theory" means language describing intersubjectively experienced mental states such as ideas and motivations.

Some romantics portray the theory-creation process as consisting firstly of introspection by the theorist upon his own personal intersubjective experiences or imagination. Then secondly it consists of the theorist imputing his introspectively experienced ideas and motives to the social members under investigation. The sociologist Max Weber (1864-1920) called this *verstehen.* When the social scientist can recognize or at least imagine the imputed ideas and motives, then the ideas and motives expressed by his theory are "convincing" to him.

3.43 Pragmatic Definition of Theory Language

Scientific theories are universally quantified language that can be schematized as nontruth-functional conditional statements including mathematical expressions (a.k.a. "models") that are proposed for empirical testing.

Unlike positivists and romantics **pragmatists define theory language pragmatically, *i.e.*, by its function in basic research**, instead of syntactically as an axiomatic system or semantically by some distinctive content. The neopragmatist definition contains the traditional idea that theories are hypotheses, but the reason for their hypothetical status is not due either to the positivist observation-theory dichotomy or to the romantics' requirement of referencing intersubjective mental states. Theory language is hypothetical because interested scientists agree that in the event of falsification, it is the theory language that is deemed falsified instead of the test-design language. Often theories are deemed to be more hypothetical, because their semantics is believed to be more empirically underdetermined than the test-design language.

Theory is a special function of language – empirical testing – rather than a special type of language.

Scientists decide that proposed theory statements are more likely to be productively revised than presumed test-design statements, if a falsifying test outcome shows that revision is needed.

After a conclusive test outcome, the tested theory is no longer a theory, because the conclusive test makes the theory either a scientific law or falsified discourse.

Pragmatically after a theory is tested, it ceases to be a theory, because it is either scientific law or rejected language, except for the skeptical scientist who can describe further predictive testing. Designing empirical tests can tax the ingenuity of the most brilliant scientist, and theories may have lives lasting many years due to difficult problems in formulating or implementing decisive test designs. Or as in a computerized discovery system with an empirical decision procedure, theories may have lives measured in milliseconds.

Romantic social scientists adamantly distinguish theory from "models". Many alternative supplemental speculations about motives, which they call "theory", can be appended to an empirical model that has been tested. But it is the model that is empirically tested statistically and/or predictively. Pragmatically the language that is proposed for empirical testing is theory, such that when a model is proposed for testing, the model has the status of theory.

Sometime after initial testing and acceptance, a scientific law may revert to theory status to be tested again. Centuries after Newton's law of gravitation had been accepted as scientific law; it was tested in 1919 in the historic Eddington eclipse test of Einstein's alternative relativity theory. Thus for a time early in the twentieth century Newton's theory was pragmatically speaking actually a theory again.

On the pragmatic definition "theory" identifies the transient status of language that is proposed for testing.

On the archival definition "theory" identifies a permanent status of accepted language as in an historical archive.

The term "theory" is ambiguous; archival and pragmatic meanings can be distinguished. In the archival sense philosophers and scientists still may speak of Newton's "theory" of gravitation, as is often done herein. The archival meaning is what in his *Patterns of Discovery* Hanson calls "completed science" or "catalogue science" as opposed to "research science". The archival sense has long-standing usage and will be in circulation for a long time to come.

The mummifying archival sense is not the meaning needed to understand the research practices and historical progress of basic science. Research scientists seeking to advance their science using theory in the archival sense instead of the functional concept are misdirected away from advancement of science. They resemble archivists and antiquarians.

Philosophers of science today recognize the pragmatic meaning of "theory", which describes it as a transitional phase in the history of science. Pragmatically Newton's "theory" is now falsified physics in basic science and is no longer proposed for testing, although it is still used by aerospace engineers and others who can exploit its lesser realism and lesser truth.

3.44 Pragmatic Definition of Test-Design Language

Pragmatically theory is universally quantified language that is <u>proposed</u> for testing, and test-design language is universally quantified language that is <u>presumed</u> for testing.

Accepting or rejecting the hypothesis that there are white crows presumes a prior agreement about the semantics needed to identify a bird's species. The test-design language defines the semantics that identifies the subject of the tested theory and the procedures for executing the test. Its semantics also includes and is not limited to the language for describing the design of any test apparatus, the testing methods including any measurement procedures, and the characterization of the theory's initial conditions. The semantics for the independent characterization of the observed outcome resulting after the test execution is also defined in the test design language. The universally quantified test-design statements contribute these meaning components to the semantics of the descriptive terms common to both the test design and the theory.

Both theory and test-design language are believed to be true, but for different reasons. Experimenters testing a theory presume the test-design language is true with definitional force for identifying the subject of the test and for executing the test design. The advocates proposing or supporting a theory believe the theory statements are true with sufficient plausibility to warrant the time, effort and cost of testing with an expected nonfalsifying test outcome. For these advocates both the theory statements and the test-design statements contribute component parts to the complex semantics of the descriptive terms that the theory and test-design statements share prior to testing. However during the test only the test-design statements have definitional force, so that the test has contingency.

Often test-design concepts describing the subject of a theory are either not yet formulated or are too vaguely described and conceptualized to be used for effective testing. They are concepts that await future scientific and technological developments that will enable formulation of an executable and decisive empirical test. Formulating a test design capable of evaluating decisively the empirical merits of a theory often requires considerable ingenuity. Eventual formulation of specific test-design language enabling an empirical test decision supplies the additional clarifying semantics that sufficiently reduces the disabling empirical underdetermination in the descriptive terms of the theory.

3.45 Pragmatic Definition of Observation Language

Observation language in science is test-design sentences that are given particular logical quantification for describing the subject of the individual test, the test procedure and its execution including the reporting of the test outcome.

After scientists have formulated and accepted a test design, the universally quantified language setting forth the design determines the semantics of its observation language. Particularly quantified language cannot define the semantics of descriptive terms. The observation language in a test is sentences or equations with particular logical quantification accepted as experimentally or otherwise experientially true and used for description, and it includes both the test-design sentences describing the initial conditions and procedures for an individual test execution and also the test-outcome sentences reporting the outcome of an executed test. This is a pragmatic concept of observation language, because it depends on the function of such language in the test. Contrary to positivists and earlier philosophers, pragmatists reject the thesis that there is any inherently or naturally observational semantics.

If a test outcome is not a falsification, then the universally quantified theory is regarded as a scientific law, and it contributes semantical components to the complex meanings associated with the descriptive terms shared with the universally quantified test-design sentences. And the nonfalsified tested theory, *i.e.,* law, when given particular quantification may be used for observational reporting. As Einstein told Heisenberg, the theory decides what the physicist can observe.

Additionally the terms in the universally quantified test-design sentences contribute their semantics to the meaning complex of the theory's terms. These semantical contributions reduce vagueness, and do not depend on the logical derivation of test-design sentences from the theory sentences. But where such derivation is possible, coherence is increased and vagueness is thereby further reduced.

Furthermore due to such a derivation test-outcome measurement values may be changed to numerical values that still fall within the range of measurement error, and the accuracy of the measurement values may be judged improved.

3.46 Observation and Test Execution

For believers in a theory, statements *predicting* test outcomes have semantics *defined in part by universally quantified theory statements* with their logical quantification made particular to describe the individual test execution.

Statements *reporting* observed test outcomes have semantics *defined by the universally quantified test-design statements* with their logical quantification made particular to describe the results of the individual test execution.

The semantics for all the language involved in a test is defined by universally quantified statements, since particularly quantified language does not define semantics. Therefore all the language needed to realize a theory's initial conditions together with the test-outcome statements have their semantics defined by the universal statements in the test design.

For the execution of the individual test event all the statements involved have their quantification changed from universal to particular. The particularly quantified statements in the test design describing the subject of the theory are also statements used for observation. The particularly quantified theory statements together with the particularly quantified test-design statements are used to produce the prediction in the test.

For a mathematically expressed theory particular logical quantification is accomplished by assigning values by measurement to implement the theory's initial conditions needed to calculate the theory's one or several prediction variables, and then by calculating the predicted numerical values.

After the test is executed, the particularly quantified statements in the test design reporting the test outcome are statements used for the semantics that describes the observed results of the test. The prediction statements are not as such observation statements unless the test outcome is nonfalsifying. If the test is falsifying, the falsified theory and its erroneous predictions are merely rejected language.

For a mathematically expressed theory a nonfalsifying test outcome is a predicted magnitude that deviates from the measurement magnitude for the same variable by an amount that is within the estimated measurement errors, such that the prediction is deemed to be as the test-outcome statements describe. Then the test is effectively decidable as nonfalsifying. Otherwise the test has falsified the theory, and the prediction values are simply rejected as erroneous.

3.47 Scientific Professions

In computational philosophy of science a "scientific profession" means the researchers who at a given point in time are attempting to solve a scientific problem as defined by a test design.

They are the language community represented by the input and output state descriptions for a discovery system application. On this definition of profession for discovery systems in computational philosophy of science, a profession is a much smaller group than the academicians in the field of the problem and is furthermore not limited to academicians.

3.48 Semantic Individuation of Theories

Theory language is *defined pragmatically*, but theories are *individuated semantically*. Theories are individuated semantically in either of two ways:

***Firstly* different expressions are different theories, because they address different subjects.** Different theory expressions having different test designs are different theories with different subjects.

***Secondly* different expressions are different theories, because each makes contrary claims about a common subject.** The test-design language defines the common subject. This is equivalent to Popper's basis for individuating theories.

Another problem with individuation of theories is the **linguistic boundary of a theory.** Theory is universally quantified language proposed for testing, and thus the choice of what language is theory and what language is test design is the result of a decision by the proposing theorist. So too, the boundary of the language of the theory having a given test design is a result of the theorist decision. The extent of language will be determined by the problem the theory addresses, and it will include all that the theorist believes is necessary to solve the problem, i.e., necessary to make an empirically adequate explanation and make empirically adequate predictions. If the theory is falsified in a test, then he may decide to incorporate more language that he believes is strategic to success, thereby practicing the discovery practice of theory elaboration, or he may call upon the other discovery practices of theory extension or theory revision.

Chapter 4. Functional Topics

The preceding Chapters have offered generic sketches of the principal twentieth-century philosophies of science, namely romanticism, positivism and pragmatism. And they have discussed selected elements of the contemporary pragmatist philosophy of language for science, namely the object language and metalanguage perspectives, the synchronic and diachronic views, and the syntactical, semantical, ontological and pragmatic dimensions.

Finally at the expense of some repetition this Chapter integrates those discussions into the four functional topics that were briefly examined in the first and second Chapters, namely the institutionalized aim of basic science, scientific discovery, scientific criticism, and scientific explanation.

4.01 Institutionalized Aim of Science

During the last approximately three hundred years empirical science has evolved into a social institution with its own distinctive and autonomous professional subculture of shared views and values.

The institutionalized aim of science is the cultural value system that regulates the scientist's performance of basic research.

Idiosyncratic motivations of individual scientists are historically noteworthy, but are largely of anecdotal interest for philosophers of science, except when such idiosyncrasies have produced results that have initiated an institutional change.

The literature of philosophy of science offers various proposals for the aim of science. The three modern philosophies of science mentioned above set forth different philosophies of language, which influence their diverse concepts of all four of the functional topics including the aim of science.

4.02 Positivist Aim

Early positivists aimed to create explanations having objective basis in observations and to make empirical generalizations summarizing the individual observations. They rejected all theories as speculative and therefore unscientific.

The positivists proposed a foundational agenda based on their naturalistic philosophy of language. Early positivists such as Mach proposed that science should aim for firm objective foundations by relying exclusively on observation, and should seek empirical generalizations that summarize the individual observations. They deemed theories to be at best temporary expedients and too speculative to be considered appropriate for science. However, Pierre Duhem (1861-1916) admitted that physical theories are integral to science, and he maintained that their function is to summarize laws as Mach said laws summarize observations, although Duhem denied that theories have a realistic semantics thus avoiding the neopositivists' problems with theoretical terms.

Later neopositivists aimed furthermore to justify explanatory theories by logically relating the theoretical terms in the theories to observation terms that they believed are a foundational reduction base.

After the acceptance of Einstein's relativity theory by physicists, the later positivists also known as "neopositivists" acknowledged the essential rôle that hypothetical theory must have in the aim of science. Between the twentieth-century World Wars, Carnap and his fellows in the Vienna Circle group of neopositivists attempted to justify theories in science by logically relating the so-called theoretical terms in the theories to the so-called observation terms that they believed should be the foundational logical-reduction base for science.

Positivists alleged the existence of "observation terms", which are terms that reference observable entities or phenomena. Observation terms are deemed to have simple, elementary and primitive semantics and to receive their semantics ostensively and passively in perception. Positivists furthermore called the particularly quantified sentences containing only such terms "observation sentences", if issued on the occasion of observing. For example the sentence "That crow is black" uttered while the speaker of the sentence is viewing a present crow, is an observation sentence.

Many of these neopositivists were also called "logical positivists", because they attempted to use the symbolic logic fabricated by Bertrand Russell (1872-1970) and Alfred N. Whitehead (1861-1949) to accomplish the logical reduction of theory language to observation language. The logical positivists fantasized that this Russellian symbolic logic can serve philosophy as mathematics serves physics, and it became their *idée fixe*. For decades the symbolic logic ostentatiously littered the pages of the *Philosophy of Science* and *British Journal for Philosophy of Science* journals with its chicken tracks, and rendered their ostensibly "technical" papers fit for the bottom of a birdcage.

These neopositivists were self-deluded, because in fact the truth-functional logic cannot capture the hypothetical-conditional logic of empirical testing in science. For example the truth-functional truth table says that if the conditional statement's antecedent statement is false, then the conditional statement expressing the theory is defined as true no matter whether the consequent is true or false. But in the practice of science a false antecedent statement means that execution of a test did not comply with the definition of initial conditions in the test design thus invalidating the test, and is therefore irrelevant to the truth-value of the conditional statement that is the tested theory. Consequently the aim of these neopositivist philosophers was not relevant to the aim of practicing research scientists.

Now the period of pretext has past. Today truth-functional logic is not seriously considered by post-positivist philosophers of science much less by practicing research scientists. Scientists do not use symbolic logic or seek any logical reduction for so-called theoretical terms. The extinction of positivism was in no small part due to the disconnect between the positivists' philosophical agenda and the actual practices and values of research scientists.

For more about positivism readers are referred to **BOOKs II** and **III** at the free web site www.philsci.com or in the e-book *Twentieth-Century Philosophy of Science*: *A History*, which is available at Internet booksellers.

4.03 Romantic Aim

The aim of the *social sciences* is to develop explanations describing social-psychological intersubjective motives, in order to explain observed social interaction in terms of purposeful "human action" in society.

The romantics have a subjectivist social-psychological reductionist aim for the social sciences, which is thus also a foundational agenda. This agenda is a thesis of the aim of the social sciences that is still enforced by many social scientists.

Thus both romantic philosophers and romantic scientists maintain that the sciences of culture differ fundamentally in their aim from the sciences of nature.

Some romantics call their type of explanation "interpretative understanding" and others call it "substantive reasoning". Using this concept of the aim of social science they often say that an explanation must be "convincing" or must "make substantive sense" to the social scientist due to the scientist's introspection upon his actual or imaginary personal experiences, especially when he is a participating member of the same culture as the social members he is investigating. Some romantics advocate "hermeneutics", which is a concept often associated with literary interpretation, and that is the hidden meaning in a text accessed by re-experiencing the inner intersubjective mental experience of a text's author.

Examples of romantic social scientists are sociologists like Talcott Parsons (1902-1979), an influential American sociologist who taught at Harvard University. In his *Structure of Social Action* (1937) he advocated a variation on the philosophy of the sociologist Max Weber, in which vicarious understanding that he called "*verstehen*" is a criterion for criticism that the romantics believe trumps empirical evidence. *Verstehen* sociology is also known as "folk sociology" or "pop sociology". Enforcing this "social action" criterion has obstructed the evolution of sociology into a modern empirical science in the twentieth century. Cultural anthropologists furthermore reject *verstehen* as a fallacy of ethnocentrism.

An example of an economist whose philosophy of science is paradigmatically romantic is Ludwig von Mises (1881-1973), an Austrian School economist. In his *Human Action: A Treatise on Economics* (1949) Mises proposes a general theory of human action that he calls "praxeology" that employs the "method of imaginary constructions", which suggests Weber's ideal types. He finds praxeology exemplified in both economics and politics. Mises maintains that praxeology is deductive and apriori like geometry, and is therefore unlike natural science. Praxeological theorems cannot be falsified, because they are certain. All that is needed for deduction of its theorems is knowledge of the "essence" of human action, which is known introspectively. On this view experience merely directs the investigator's interest to problems.

The 1989 Nobel-laureate econometrician Trygve Haavelmo (1911-1999) in his "Probability Approach in Econometrics" in *Econometrica* (July supplement, 1944) supplies another far more widely accepted example of romanticism. These econometricians do not reject the aim of prediction, simulation, optimization and policy formulation using statistical econometric models; with their econometric modeling agenda they enable it. But they subordinate the selection of explanatory variables in their models to factors that are derived from economists' heroically imputed maximizing rationality theses, which identify the motivating factors explaining the decisions of economic agents such as buyers and sellers in a market. Thus they exclude econometrics from discovery and limit its function to testing romantic "theory". In his *Philosophy of Social Science* (1995) Alexander Rosenberg (1946) describes the economists' theory of "rational choice", *i.e.*, the use of the maximizing rationality theses, as "folk psychology formalized".

However the "theoretical" economist's rationality postulates have been relegated to the status of a fatuous *cliché*, because in practice the econometrician almost never derives his equation specification deductively from the rationality postulates expressed as preference schedules. Instead he will select variables to produce statistically good models regardless of the rationality postulates. In fact in Haavelmo's seminal paper he wrote that the economist may "jump over the middle link" of the preference schedules, although he rejected determining equation specifications by statistics alone.

For more about the romantics including Parsons, Weber, Haavelmo and others readers are referred to **BOOK VIII** at the free web site www.philsci.com or in the e-book *Twentieth-Century Philosophy of Science: A History*, which is available from most Internet booksellers.

4.04 More Recent Ideas

Most of the twentieth-century philosophers' proposals for the aim of science are less dogmatic than those listed above and arise from examination of important developmental episodes in the history of the natural sciences. Some noteworthy examples:

Einstein: Reflection on his relativity theory influenced Albert Einstein's concept of the aim of science, which he set forth as his "programmatic aim of all physics" stated in his "Reply to Criticisms" in *Albert Einstein: Philosopher-Scientist* (1949). The aim of science in Einstein's view is a comprehension as complete as possible of the connections among sense impressions in their totality, and the accomplishment of this comprehension by the use of a minimum of primary concepts and relations. Einstein certainly did not reject empiricism, but he included an explicit coherence agenda in his aim of science. This thesis also implies a uniform ontology for physics, and Einstein accordingly found statistical quantum theory to be "incomplete" according to his aim, which is a minority view today.

Popper: Karl R. Popper was an early post-positivist philosopher of science and was also critical of the romantics. Reflecting on Sir Arthur Eddington's (1882-1944) historic 1919 solar eclipse test of Einstein's relativity theory in physics Popper proposed in his *Logic of Scientific Discovery* (1934) that the aim of science is to produce tested and nonfalsified theories having greater universality and more information content than any predecessor theories addressing the same subject. Unlike the positivists' view his concept of the aim of science thus focuses on the growth of scientific knowledge. And in his *Realism and the Aim of Science* (1983) he maintains that realism explains the possibility of falsifying test outcomes in scientific criticism. The title of his *Logic of Scientific Discovery* notwithstanding, Popper denies that discovery can be addressed by either logic or philosophy, but says instead that discovery is a proper subject for psychology. Cognitive psychologists today would agree.

Hanson: Norwood Russell Hanson reflecting on the development of quantum theory states in his *Patterns of Discovery: An Inquiry into the Conceptual Foundations of Science* (1958) and in *Perception and Discovery: An Introduction to Scientific Inquiry* (1969) that inquiry in research science is directed to the discovery of new patterns in data to develop new hypotheses for deductive explanation. He calls such practices "research science", which he opposes to "completed science" or "catalogue science", which is merely re-arranging established ideas into more elegant formal axiomatic patterns. He follows Peirce who called hypothesis formation "abduction". Today mechanized discovery systems typically search for patterns in data.

Kuhn: Thomas S. Kuhn, reflecting on the development of the Copernican heliocentric cosmology in his *The Copernican Revolution: Planetary Astronomy in the Development of Western Thought* (1957) maintained in his popular *Structure of Scientific Revolutions* (1962) that the prevailing theory, which he called the "consensus paradigm", has institutional status. He proposed that small incremental changes extending the consensus paradigm, to which scientists seek to conform, defines the institutionalized aim of science, which he called "normal science". And he said that scientists neither desire nor aim consciously to produce revolutionary new theories, which he called "extraordinary science." This concept of the aim of science is thus a conformist agenda; Kuhn therefore defined scientific revolutions as institutional changes in science, which he excludes from the institutionalized aim of science.

Feyerabend: Paul K. Feyerabend reflecting on the development of quantum theory in his *Against Method* proposed that each scientist has his own aim, and that anything institutional is a conformist impediment to the advancement of science. He said that historically successful scientists always "break the rules", and he ridiculed Popper's view of the aim of science calling it "ratiomania" and "law-and-order science". Therefore Feyerabend proposes that successful science is literally "anarchical", and borrowing a slogan from the Marxist Leon Trotsky, Feyerabend advocates "revolution in permanence".

For more about the philosophies of Popper, Kuhn, Hanson and Feyerabend readers are referred to **BOOK**s **V, VI** and **VII** at the free web site www.philsci.com or in the e-book *Twentieth-Century Philosophy of Science*: *A History*, which is available from most Internet booksellers.

4.05 Aim of Maximizing "Explanatory Coherence"

Thagard: Computational philosopher of science Paul Thagard proposes that the aim of science is "best explanation". The thesis refers to an explanation that aims to maximize the explanatory coherence of one's overall set of beliefs. This aim of science is thus explicitly a coherence agenda.

Thagard developed a computerized cognitive system **ECHO,** an acronym for "Explanatory Coherence by Harmony Optimization", in order to explore the operative criteria in theory choice. His system described in his *Conceptual Revolutions* (1992) simulated the realization of the aim of maximizing "explanatory coherence" by replicating various episodes of theory choice in the history of science. In his system "explanation" is an undefined primitive term. He applied his system **ECHO** to replicate theory choices in several episodes in the history of science including (1) Lavoisier's oxygen theory of combustion, (2) Darwin's theory of the evolution of species, (3) Copernicus' heliocentric astronomical theory of the planets, (4) Newton's theory of gravitation, and (5) Hess' geological theory of plate tectonics.

In reviewing his historical simulations Thagard reports that **ECHO** indicates that the criterion making the largest contribution historically to explanatory coherence in scientific revolutions is explanatory breadth – the preference for the theory that explains more evidence than its competitors. But he adds that the simplicity and analogy criteria are also historically operative although less important. He maintains that the aim of maximizing explanatory coherence with these three criteria yields the "best explanation".

"Explanationism", maximizing the explanatory coherence of one's overall set of beliefs, is inherently conservative. The **ECHO** system appears to document the historical fact that the coherence aim is psychologically satisfying and occasions strong, and for some scientists nearly compelling motivation for accepting coherent theories, while theories describing reality as incoherent with established beliefs are psychologically disturbing, and are often rejected when first proposed. But progress in science does not consist in maximizing the scientist's psychological contentment. Empiricism eventually overrides coherence when there is a conflict due to new evidence. In fact defending coherence has historically had a reactionary effect. For example Heisenberg's revolutionary indeterminacy relations, which contradict microphysical theories coherent with established classical physics including Einstein's general relativity theory, do not conform to **ECHO**'s maximizing-explanatory-coherence criterion.

For more about the philosophy of Thagard readers are referred to **BOOK VIII** at the free web site www.philsci.com or in the e-book *Twentieth-Century Philosophy of Science: A History*, which is available from most Internet booksellers.

4.06 Contemporary Pragmatist Aim

The successful outcome (and thus the aim) of basic-science research is explanations made by developing theories that satisfy critically empirical tests, which theories are thereby made scientific laws that can function in scientific explanations and test designs.

The principles of contemporary pragmatism including its philosophy of language evolved through the twentieth century beginning with the autobiographical writings of Heisenberg, one of the central participants in the historic development of quantum theory. This philosophy is summarized in Section **2.03** above in three central theses: relativized semantics, empirical underdetermination and ontological relativity, which are not repeated here.

For more about the philosophy of Heisenberg readers are referred to **BOOKs II** and **IV** at the free web site www.philsci.com or in the e-book *Twentieth-Century*

Philosophy of Science: *A History*, which is available from most Internet booksellers.

The institutionally regulated practices of research scientists may be described succinctly in the pragmatist statement of the aim of science. The contemporary research scientist seeking success in his research may consciously employ this aim as what some social scientists call a "rationality postulate". The institutionalized aim of science can be expressed as such a pragmatist "rationality postulate" as follows:

The institutionalized aim of science is to construct explanations by developing theories that satisfy critically empirical tests, which theories are thereby made scientific laws that can function in scientific explanations and test designs.

Pragmatically rationality is not some incorrigible principle or intuitive preconception. The contemporary pragmatist statement of the aim of science is a postulate in the sense of an empirical hypothesis about what has been and will be responsible for the historical advancement of basic-research science. Therefore like any hypothesis it is destined to be revised at some unforeseeable future time, when due to some future developmental episode in basic science, research practices are revised in some fundamental way. Then some conventional practices deemed rational today might be dismissed by philosophers and scientists as misconceptions and perhaps even superstitions as are the romantic and positivist beliefs today. The aim of science is more elaborately explained in terms of all four of the functional topics as sequential steps in the development of explanations.

The institutionalized aim can also be expressed so as not to impute motives to the successful scientist, whose personal psychological motives may be quite idiosyncratic and even irrelevant. Thus the contemporary pragmatist statement of the aim of science may instead be phrased as follows in terms of a successful outcome instead of a conscious aim imputed to scientists:

The successful outcome of basic-science research is explanations made by developing theories that satisfy critically empirical tests, which theories are thereby made scientific laws that can function in scientific explanations and test designs.

The empirical criterion is the ***only*** criterion acknowledged by the contemporary pragmatist, because it is the only criterion that accounts for the advancement of science. Historically there have been other criteria, but whenever there has been a conflict, eventually it is demonstrably superior empirical adequacy often exhibited in practicality that has enabled a new theory to prevail. This is true even if the superior theory's ascendancy has taken many years or decades, or even if it has had to be rediscovered, such as the heliocentric theory of the ancient Greek astronomer Aristarchus of Samos (in the third century BCE).

4.07 Institutional Change

Change within the institution of science is change made under the regulation of the institutionalized aim of science, and may consist of new theories, new test designs, new laws and/or new explanations.

Change of the institution of science, *i.e.*, institutional change, on the other hand is the historical evolution of scientific practices involving revision of the aim of science, which may be due to revision of its criteria for criticism, its discovery practices, or its concept of explanation.

Institutional change in science must be distinguished from change within the institutional constraint. Philosophy of science examines both changes within the institution of science and historical changes of the institution itself. But institutional change is often recognized only retrospectively due to the distinctively historical uniqueness of each episode and also due to the need for eventual conventionality for new basic-research practices to become institutionalized. The emergence of artificial intelligence in the sciences may exemplify an institutional change in progress today.

In the history of science institutionally deviate practices that yielded successful results are initially recognized and accepted by only a few scientists. As Feyerabend emphasized in his *Against Method*, in the history of science successful scientists have often broken the prevailing methodological rules. But the successful departures eventually become conventionalized. And that is clearly true of the quantum theory. By the time they are deemed acceptable to the peer-reviewed literature, reference manuals, encyclopedias, student textbooks and desperate academic plagiarists, the institutional change is complete and has become the received conventional wisdom.

Successful researchers have often failed to understand the reasons for their unconventional successes, and have advanced or accepted erroneous methodological ideas and philosophies of science to explain their successes. One of the most historically notorious such misunderstandings is Isaac Newton's "*hypotheses non fingo*", his denial that his law of gravitation is a hypothesis. Nearly three centuries later Einstein demonstrated otherwise.

Newton's physics occasioned an institutional change in physicists' concept of explanation. Newton's contemporaries, Gottfried Leibniz (1646-1716) and Christian Huygens

(1629-1695) had criticized Newton's gravitational theory for admitting action at a distance. Both of these contemporaries of Newton were convinced that all physical change must occur through direct physical contact like colliding billiard balls. Leibniz therefore described Newton's concept of gravity as an "occult quantity" and called Newton's theory unintelligible. But eventually Newtonian mathematical physics became institutionalized and paradigmatic of explanation in physics. For example in the late nineteenth century the physicist Hermann von Helmholtz (1821-1894) said that to understand a phenomenon in physics means to reduce it to Newtonian laws.

In his *Concept of the Positron* (1963) Hanson proposes three stages in the process of the evolution of a new concept of explanation; he calls them the black box, the gray-box, and the glass box stages. In the initial black-box stage, there is an algorithmic novelty, a new formalism, which is able to account for all the phenomena that an existing formalism can account for. Scientists use this technique, but they then attempt to translate its results into the more familiar terms of the prevailing orthodoxy, in order to provide "understanding". In the second stage, the gray-box stage, the new formalism makes superior predictions in comparison to older alternatives, but it is still viewed as offering no "understanding". Nonetheless it is suspected as having some structure that is in common with the reality it predicts. In the final glass-box stage the success of the new theory will have so permeated the operation and techniques of the body of the science that its structure will also appear as the proper pattern of scientific inquiry.

Einstein was never able to accept the Copenhagen statistical interpretation and a few physicists today still reject it. Writing in 1958 Hanson said that quantum theory is in the gray-box stage, because scientists have not yet ceased to distinguish between the theory's structure and that of the phenomena themselves. This is to say that they did not practice ontological relativity. But since Aspect, Dalibard, and Roger's findings from their 1982 nonlocality experiments demonstrated empirically the Copenhagen interpretation's semantics and ontology, the quantum theory-based evolution of the concept of explanation in physics has become institutionalized.

4.08 Philosophy's Cultural Lag

Recognition of successful departures from institutionalized basic research is elusive even for philosophers. **There exists a time lag between the evolution of the institution of science and developments in philosophy of science, since the latter depend on the realization of the former** and since philosophers tend to be more doctrinaire than scientists who use their empirical criteria. For example more than a quarter of a century passed between Heisenberg's philosophical reflections on the language of his indeterminacy relations in quantum physics and the emergence of the contemporary pragmatist philosophy of science in academic philosophy. Heisenberg is one of the twentieth century's great philosophers of language. But even today academic philosophers almost never reference his philosophical writings, because he is uncomprehendingly dismissed as merely a physicist and not one of them; he is still ignored by many of them as just a pretentiously meddling layman in academic philosophy.

4.09 Cultural Lags among Sciences

Not only are there cultural lags between the institutionalized practices of science and philosophy of science, **there are also cultural lags among the several sciences.**

Philosophers of science have preferred to examine physics and astronomy, because historically these have been the most advanced sciences since the historic Scientific

Revolution benchmarked with Copernicus and Newton. Institutional changes occur with lengthy time lags due to such impediments as intellectual mediocrity, technical incompetence, risk aversion, or vested interests in the conventional ideas of the received wisdom. As Planck grimly wrote in his *Scientific Autobiography* (1949), a new truth does not triumph by convincing its opponents, but rather succeeds because its opponents have died off; or as he also said, science progresses "funeral by funeral".

The newer social and behavioral sciences have remained institutionally retarded. Naïve sociologists and economists even today are blithely complacent in their amateurish philosophizing about basic social-science research, often adopting prescriptions and proscriptions that contemporary philosophers of science recognize as anachronistic and counterproductive. The result has been the emergence and survival of retarding philosophical superstitions in these retarded social sciences, especially to the extent that they have looked to their own less successful histories to formulate their ersatz philosophies of science.

Currently most sociologists and economists still enforce a romantic philosophy of science, because they believe that sociocultural sciences must have fundamentally different philosophies of science than the natural sciences. Similarly behaviorist psychologists continue to impose the anachronistic positivist philosophy of science. On the contemporary pragmatist philosophy these sciences are institutionally retarded, because they erroneously impose preconceived semantical and ontological commitments as criteria for scientific criticism.

Pragmatists can agree with Popper, who in his critique of Kuhn in "Normal Science and its Dangers" in *Criticism and the Growth of Knowledge* (1970) said that science is "subjectless" meaning that valid science is not defined by any particular semantics or ontology. Pragmatists tolerate any semantics or ontology that romantics or positivists may include in scientific explanations, theories and laws, but pragmatists recognize ***only*** the empirical criterion for criticism.

4.10 Scientific Discovery

"Discovery" refers to the development of new and empirically superior theories.

Much has already been said in the above discussions of philosophy of scientific language in Chapter 3 about the pragmatic basis for the definition of theory language, about the semantic basis for the individuation of theories, and about state descriptions. Those discussions will be assumed in the following comments about the mechanized development of new theories.

Discovery is the first step toward realizing the aim of science. The problem of scientific discovery for contemporary pragmatist philosophers of science is to proceduralize and then to mechanize the development of universally quantified statements for empirical testing with nonfalsifying test outcomes, thereby making laws for use in explanations and test designs. **Contemporary pragmatism is consistent with the techniques of computerized discovery systems.**

4.11 Discovery Systems

A discovery system produces a transition from an input-language state description containing currently available language to an output-language state description containing generated and tested new theories.

In the "Introduction" to his *Models of Discovery* (1977) Simon, one of the founders of artificial intelligence wrote that dense mists of romanticism and downright know-nothingism have always surrounded the subject of scientific discovery and creativity. Therefore the most significant development addressing the problem of scientific discovery has been the relatively recent mechanized discovery systems in a new specialty called "computational philosophy of science".

The ultimate aim of the computational philosopher of science is to facilitate the advancement of contemporary sciences by participating in and contributing to the successful basic-research work of the scientist. The contemporary pragmatist philosophy of science thus carries forward the classical pragmatist John Dewey's emphasis on participation. Unfortunately few academic philosophers have the requisite computer skills much less a working knowledge of any empirical science for participation in basic research. Hopefully that will change in future Ph.D. dissertations in philosophy of science, which are likely to be interdisciplinary endeavors.

Every useful discovery system to date has contained procedures both for constructional theory creation and for critical theory evaluation for quality control of the generated output and for quantity control of the system's otherwise unmanageably large output. Theory creation introduces new language into the current state description to produce a new state description, while falsification in empirical tests eliminates language from the current state description to produce a new state description. Thus both theory development and theory testing enable a discovery system to offer a specific and productive diachronic dynamic procedure for linguistic change to advance empirical science.

The discovery systems do not merely implement an inductivist strategy of searching for repetitions of individual instances, notwithstanding that statistical inference is employed in some system designs. The system designs are mechanized procedural strategies that search for patterns in the input information. Thus they implement Hanson's thesis in *Patterns of Discovery* that in a growing research discipline inquiry seeks the discovery of new patterns in data. They also implement Feyerabend's "plea for hedonism" in *Criticism and the Growth of Knowledge* (1971) to produce a proliferation of theories. But while many are made by these systems, mercifully few are chosen thanks to the empirical testing routines in the systems to control for quality of the outputted equations.

4.12 Types of Theory Development

In his *Introduction to Metascience* Hickey distinguishes three types of theory development, which he calls **theory extension**, **theory elaboration** and **theory revision.** This classification is vague and may be overlapping in some cases, but it suggests three alternative types of discovery strategies and therefore implies different discovery-system designs.

***Theory extension* is the use of a currently tested and nonfalsified explanation to address a new scientific problem.**

The extension could be as simple as adding hypothetical statements to make a general explanation more specific for a new problem at hand. A more complex strategy for theory extension is analogy. In his *Computational Philosophy of Science* (1988) Thagard describes this strategy for mechanized theory development, which consists in the patterning of a proposed solution to a new problem by analogy with a successful explanation originally developed for a different subject. Using his system design based on this strategy his discovery system called **PI** (an acronym for "Process of Induction") reconstructed development of the theory of sound waves by analogy with the description of water waves. The system was his Ph.D. dissertation in philosophy of science at the University of Toronto, Canada.

In his *Mental Leaps*: *Analogy in Creative Thought* (1995) Thagard further explains that analogy is a kind of nondeductive logic, which he calls "analogic". It firstly involves the "source analogue", which is the known domain that the investigator already understands in terms of familiar patterns, and secondly involves the "target analogue", which is the unfamiliar domain that the investigator is trying to explain. Analogic is the strategy whereby the investigator understands the targeted domain by seeing it in terms of the source domain. Analogic requires a "mental leap", because the two analogues may initially seem unrelated. And the mental leap is also called a "leap", because analogic is not conclusive like deduction.

It may be noted that if the output state description generated by analogy such as the **PI** system is radically different from anything previously seen by the affected scientific profession containing the target analogue, then the members of that affected profession may experience the communication constraint to the high degree that is usually associated with a theory revision. The communication constraint is discussed below (See below, Section **4.26**).

Theory elaboration **is the correction of a currently falsified theory to create a new theory by adding new factors or variables that correct the falsified universally quantified statements and erroneous predictions of the old theory.**

The new theory has the same test design as the old theory. The correction is not merely *ad hoc* excluding individual exceptional cases, but rather is a change in the universally quantified statements. This process is often misrepresented as "saving" a falsified theory, but in fact it creates a new one.

For example the introduction of a variable for the volume quantity and development of a constant coefficient for the particular gas could elaborate Gay-Lussac's law for gasses into the combined Gay-Lussac's law, Boyle's law and Charles' law. Similarly Friedman's macroeconomic quantity theory might be elaborated into a Keynesian hyperbolic liquidity-preference function by the introduction of an interest rate, both to account for the cyclicality manifest in an annual time series describing the calculated velocity parameter and to display the liquidity trap phenomenon, which occurred both in the Great Depression (1929-1933) and in the recent Great Recession (2007-2009).

Pat Langley's **BACON** discovery system exemplifies mechanized theory elaboration. It is named after the English philosopher Francis Bacon (1561-1626) who thought that scientific discovery can be routinized. **BACON** is a set of successive and increasingly sophisticated discovery systems that make quantitative laws and theories from input measurements. Langley designed and implemented **BACON** in 1979 as the thesis for his Ph.D. dissertation written in the Carnegie-Mellon department of psychology under the direction of Simon. A description of the system is in Simon's *Scientific Discovery*: *Computational Explorations of the Creative Processes* (1987).

BACON uses Simon's heuristic-search design strategy, which may be construed as a sequential application of theory elaboration. Given sets of observation measurements for several variables, **BACON** searches for functional relations among the variables. Langley reports that **BACON** has simulated the discovery of several historically significant empirical laws including Boyle's law of gases, Kepler's third planetary law, Galileo's law of motion of objects on inclined planes, and Ohm's law of electrical current.

Theory revision **is the reorganization of currently available information to create a new theory.**

The results of theory revision may be radically different from any current theory, and may thus be said to occasion a "paradigm change". It might be undertaken after repeated attempts at both theory extension and theory elaborations have failed. The source for the input state description for mechanized theory revision presumably consists of the descriptive vocabulary from any currently untested theories addressing the problem at hand. The descriptive vocabulary from previously falsified theories may also be included as inputs to make an accumulative state description, because the vocabularies in rejected theories can be productively cannibalized for their scrap value. In fact even terms and variables from tested and nonfalsified theories could also be included, just to see what new proposals come out; empirical underdetermination permits scientific pluralism, and reality is full of surprises. Hickey notes that a mechanized discovery system's newly outputted theory is most likely to be called revolutionary if the revision is great, because theory revision typically produces greater change to the current language state than does theory extension or theory elaboration thus producing psychologically disorienting semantical dissolution in the transition.

Theory revision, the reorganization of currently existing information to create a new theory, is evident in the history of science. The central thesis of historian of science Herbert Butterfield's (1900-1979) *Origins of Modern Science: 1300-1800* (1958, P. 1) is that the type of transition known as a "scientific revolution" was not brought about by new observations or additional evidence, but rather by transpositions in the minds of the scientists. Specifically he maintains that the type of mental activity that produced the historic scientific revolutions is the art of placing a known bundle of data in a new system of relations.

Hickey found this "art" in the history of economics. The applicability of his theory-revising **METAMODEL** discovery system to the development of Keynes' general theory was already known in retrospect by the fact that, as 1980 Nobel-laureate econometrician Lawrence Klein wrote in his *Keynesian Revolution* (1949, Pp. 13 & 124), all the important parts of Keynes theory can be found in the works of one or another of his predecessors. Thus Keynes put a known bundle of information into a new system of relations, such as his aggregate consumption function and his demand for money with its speculative-demand component with its liquidity trap.

Hickey's **METAMODEL** discovery system constructed the Keynesian macroeconomic theory from U.S. statistical data available prior to 1936, the publication year of Keynes' revolutionary *General Theory of Employment, Interest and Money*. Hickey's **METAMODEL** discovery system described in his *Introduction to Metascience* (1976) is a mechanized generative grammar with combinatorial transition rules for producing longitudinal econometric models. Hickey's mechanized grammar is a combinatorial finite-state generative grammar that satisfies the collinearity restraint for the regression-estimated equations and for the formal requirements for executable multi-equation predictive models. The system tests for collinearity, statistical significance, serial correlation, goodness-of-fit and for accurate out-of-sample retrodictions.

Hickey also used his **METAMODEL** system in 1976 to develop a small post-classical macrosociometric functionalist model of the American national society with fifty years of historical time-series data. The generated sociological model disclosed an ***intergenerational negative feedback*** that sociologists would call a "macrosocial integrative mechanism", in which an increase in social disorder indicated by a rising homicide rate calls forth a delayed intergenerational stabilizing reaction by the socializing institution indicated by the high school completion rate, which tends to restore order by reinforcing compliance with criminal law. To the shock, chagrin and dismay of complacent academic sociologists it is not a social-psychological theory, and four sociological journals therefore rejected Hickey's paper, which describes the model and its findings about the American national society's dynamics and stability characteristics. The paper is reprinted as **"Appendix I"** to **BOOK VIII** at the free web site www.philsci.com or in the e-book *Twentieth-Century Philosophy of Science*: *A History*.

The provincial academic sociologists' a priori ontological commitments to romanticism and to social-psychological reductionism rendered the editors and their chosen referees invincibly obdurate. In their criticisms cited as reasons for rejection of Hickey's paper the editors' favorite referees also exhibited their Luddite mentality toward mechanized theory development. The referee criticisms and Hickey's rejoinders are given in **"Appendix II"** to **BOOK VIII** at the free web site www.philsci.com and in the e-book *Twentieth-Century Philosophy of Science*: *A History*. These documents reveal why contemporary academic macrosociology is a stagnant intellectual backwater. But With Hickey's luck some academic sociologist will plagiarize Hickey's copyrighted findings, and then find an editor of an academic sociology journal who will publish it.

Simon called the combinatorial system design a "generate-and-test" design. In the 1980's he had written that combinatorial procedures consume excessive computational resources for present-day electronic computers, but developments in quantum computing will overcome such constraints, where the constraints are currently encountered. The increase in throughput enabled by the quantum computer is extraordinary relative to the conventional electronic computer – even the supercomputer. And the availability of practical quantum computing seems only a matter of time. Google's Research Lab in Santa Barbara, CA, recently announced in the scientific journal *Nature* that its computer scientists have achieved "quantum supremacy". *The New York Times* (24 October 2019) quoted Dr. John Martinis, project leader for Google's "quantum supremacy experiment" as saying that his group is now at the stage of trying to make use of this enhanced computing power. The article also quoted Dr. Dario Gill of IBM as predicting that by 2020 quantum computing will be in use for commercial and scientific advantage.

In the mid-1980's Hickey integrated his macrosociometric model into a Keynesian macroeconometric model to produce an institutionalist macroeconometric model, while he was employed as Deputy Director and Senior Economist for the Indiana Department of Commerce, Division of Economic Analysis during the Orr-Mutz Administration (1981-1988). The report of the findings was read to the Indiana Legislative Assembly by the Speaker of the House in support of Governor Orr's "A-plus" successful legislative initiative for increased State-government spending for primary and secondary public education.

4.13 Examples of Successful Discovery Systems

There are several examples of successful discovery systems in use. John Sonquist developed his **AID** system for his Ph.D. dissertation in sociology at the University of Chicago. His dissertation was written in 1961 before Edward O. Laumann and the romantics took over the University of

Chicago sociology department. He described the system in his *Multivariate Model Building: Validation of a Search Strategy* (1970). The system has long been used at the Survey Research Center, Institute for Social Research, University of Michigan, Ann Arbor, MI. Now modified as the **CHAID** system using chi-squared (χ^2) Sonquist's discovery system is available commercially in both the SAS and SPSS software packages. Its principal commercial application is for list-processing scoring models for commercial market analysis and for credit risk analysis as well as for academic investigations in social science. It is not only the oldest mechanized discovery system but also the most widely used in practical applications to date.

Robert Litterman developed his **BVAR** (Bayesian Vector Autoregression) system for his Ph.D. dissertation in economics at the University of Minnesota. He described the system in his *Techniques for Forecasting Using Vector Autoregressions* (1984). The economists at the Federal Reserve Bank of Minneapolis have used his system for macroeconomic and regional economic analysis. The State of Connecticut and the State of Indiana have also used it for regional economic analysis.

Having previously received an M.A. degree in economics Hickey had intended to develop his **METAMODEL** computerized discovery system for a Ph.D. dissertation in philosophy of science in the philosophy department of the University of Notre Dame, South Bend, Indiana. But Notre Dame is a Roman Catholic school with an intolerant philosophy faculty that was obstructionist to Hickey's views; their Department's Reverend Chairman told Hickey to get reformed or get out. Hickey therefore dropped out without a doctorate, and developed his computerized discovery system as a nondegree student at San Jose City College in San Jose, CA, a two-year associate-arts degree community college, with a better computer and faculty than Notre Dame.

For the next thirty years Hickey used his discovery system occupationally, working as a research econometrician in both business and government. For six of those years he used his system for institutionalist macroeconometric modeling and regional econometric modeling as Deputy Research Director and Senior Economist for the State of Indiana Department of Commerce. He also used it successfully for econometric market analysis and risk analysis for various business corporations including USX/United States Steel Corporation, BAT(UK)/Brown and Williamson Company, Pepsi/Quaker Oats Company, Altria/Kraft Foods Company, Allstate Insurance Company, and TransUnion LLC. In 2004 TransUnion's Analytical Services Group purchased a perpetual license to use his **METAMODEL** system for their consumer credit risk analyses using their proprietary TrenData aggregated quarterly time series extracted from their national database of consumer credit files. Hickey used the models generated by the discovery system to forecast payment delinquency rates, bankruptcy filings, average balances and other consumer borrower characteristics that affect risk exposure for lenders. And he used his system for Quaker Oats and Kraft Foods to discover the sociological and demographic factors responsible for the secular long-term market dynamics of processed food products and other nondurable consumer goods.

In 2007 Michael Schmidt, a Ph.D. student in computational biology at Cornell University, and his dissertation director, Hod Lipson developed their system **EUREQA** at Cornell University's Artificial Intelligence Lab. The system automatically develops predictive analytical models from data using a strategy they call an "evolutionary search" to find invariant relationships, which converges on the simplest and most accurate equations fitting the inputted data.

The system has been used by many business corporations, universities and government agencies including Alcoa, California Institute of Technology, Cargill, Corning, Dow Chemical, General Electric, Amazon, Shell and NASA.

For more about discovery systems and computational philosophy of science readers are referred to **BOOK VIII** at the free web site www.philsci.com or in the e-book *Twentieth-Century Philosophy of Science*: *A History*, which is available from most Internet booksellers.

4.14 Scientific Criticism

Criticism pertains to the criteria for the acceptance or rejection of theories. The *only* criterion for scientific criticism that is acknowledged by the contemporary pragmatist is the *empirical criterion*.

The philosophical literature on scientific criticism has little to say about the specifics of experimental design, as might be found in various college-level science laboratory manuals. Most often philosophical discussion of criticism pertains to the criteria for acceptance or rejection of theories and more recently to the effective decidability of empirical testing.

In earlier times when the natural sciences were called "natural philosophy" and social sciences were called "moral philosophy", nonempirical considerations operated as criteria for the criticism and acceptance of descriptive narratives. Even today some philosophers and scientists have used their semantical and ontological preconceptions as criteria for the criticism of theories including preconceptions about causality or specific causal factors. Such semantical and ontological preconceptions have misled them to reject new empirically superior theories. In his *Against Method* Feyerabend noted that the ontological preconceptions used to criticize new theories have often been the semantical and ontological claims expressed by previously accepted and since falsified theories.

What historically has separated the empirical sciences from their origins in natural and moral philosophy is the empirical criterion. This criterion is responsible for the advancement of science and for its enabling practicality in application. Whenever in the history of science there has been a conflict between the empirical criterion and any nonempirical criteria for the evaluation of new theories, it is eventually the empirical criterion that ultimately decides theory selection.

Contemporary pragmatists accept relativized semantics, scientific realism, and thus ontological relativity, and they therefore reject all prior semantical or ontological criteria for scientific criticism including the romantics' mentalistic ontology requiring social-psychological or any other kind of reductionism.

4.15 Logic of Empirical Testing

The logic for the rational reconstruction of scientific criticism is as follows:

An empirical test is:

(1) an effective decision procedure that can be schematized as a *modus tollens* logical deduction from a set of one or several universally quantified theory statements expressible in a nontruth-functional hypothetical-conditional schema

(2) together with a particularly quantified antecedent description of the initial test conditions as defined in the test design

(3) that jointly conclude to a consequent particularly quantified description of a produced (predicted) test-outcome event

(4) that is compared with the observed test-outcome description.

In order to express explicitly the dependency of the produced effect upon the realized initial conditions in an empirical test, the universally quantified theory statements can be schematized as a nontruth-functional hypothetical-conditional schema, *i.e.*, as a statement with the logical form "For every A if A, then C."

This hypothetical-conditional schema "For every A if A, then C." represents a system of one or several universally quantified related theory statements or equations that describe a dependency of the occurrence of events described by "C" upon the occurrence of events described by "A". In some cases the dependency is expressed as a bounded stochastic density function for the values of predicted probabilities. For advocates who believe in the theory, the hypothetical-conditional schema is the theory-language context that contributes meaning parts to the complex semantics of the theory's constituent descriptive terms including the terms common to the theory and test design. But the theory's semantical contribution cannot be operative in a test for the test to be independent of the theory, since the test outcome is not true by definition; it is empirically contingent.

The antecedent "A" includes the set of universally quantified statements of test design that describe the initial conditions that must be realized for execution of an empirical test of the theory including the statements describing the procedures needed for their realization. These statements constituting "A" are always presumed to be true or the test design is rejected as invalid, as is any test made with it. The test-design statements are semantical rules that contribute meaning parts to the complex semantics of the terms common to theory and test design, and do so independently of the theory's semantical contributions. The universal logical quantification indicates that any execution of the test is but one of an indefinitely large number of possible test executions, whether or not the test is repeatable at will.

When the test is executed, the logical quantification of "A" is changed from universal to particular quantification to describe the realized initial conditions in the individual test execution. When the universally quantified test-design and test-outcome statements have their logical quantification changed to particular quantification, the belief status and thus definitional rôle of the universally quantified test-design confer upon their particularly quantified versions the status of "fact" for all who decided to accept the test design. The theory statements in the hypothetical-conditional schema are also given particular quantification for the test execution. In a mathematically expressed theory the test execution consists in measurement actions and assignment of the resulting measurement values to the variables in "A". In a mathematically expressed single-equation theory, "A" includes the independent variables in the equation of the theory and the test procedure. In a multi-equation system whether recursively structured or simultaneous, all the exogenous variables are assigned values by measurement, and are included in "A". In longitudinal models with dated variables the lagged-values of endogenous variables that are the initial condition for a test and that initiate the recursion through successive iterations to generate predictions, must also be in "A".

The consequent "C" represents the set of universally quantified statements of the theory that correctly predict the outcome of every correct execution of a test design. Its logical quantification is changed from universal to particular quantification to describe the predicted outcome for the individual test execution. In a mathematically expressed single-equation theory the dependent variable of the theory's equation is in "C". When no value is assigned to any variable, the equation is universally quantified. When the predicted value of a dependent variable is calculated from the measurement values of the independent variables, it is particularly quantified. In a multi-equation theory, whether recursively structured or a simultaneous-equation system, the solution values for all the endogenous variables are included in "C". In longitudinal models with dated variables the current-dated values of endogenous variables for each iteration of the model, which are calculated by solving the model through successive iterations, are included in "C".

The conditional statement of theory does not say "For every A and for every C if A, then C". It only says "For every A if A, then C". In other words the conditional statement of theory only expresses a sufficient condition for the correct prediction made in C upon realization of the test conditions described in "A", and not a necessary condition. This occurs if scientific pluralism (See below, Section **4.20**) occasions multiple theories proposing alternative causal factors for the same outcome predicted correctly in "C". Or if there are equivalent measurement procedures or instruments described in "A" that produce alternative measurements with each having values falling within the range of the other's measurement error.

Let another particularly quantified statement denoted "O" describe the observed test outcome of an individual test execution. The report of the test outcome "O" shares vocabulary with the prediction statements in "C". But the semantics of the terms in "O" is determined exclusively by the universally quantified test-design statements rather than by the statements of the theory, and thus for the test its semantics is independent of the theory's semantical contribution. In an individual test execution "O" represents observations and/or measurements made and measurement values assigned apart from the prediction in "C", and it too has particular logical quantification to describe the observed outcome resulting from the individual execution of the test. There are three possible outcome scenarios:

Scenario I: If "A" is false in an individual test execution, then regardless of the truth of "C" the test execution is simply invalid due to a scientist's failure to comply with the agreed test design, and the empirical adequacy of the theory remains unaffected and unknown. The empirical test is conclusive only if it is executed in accordance with its test design. Contrary to the logical positivists, the truth table for the truth-functional logic is therefore not applicable to testing in empirical science, because in science a false antecedent, "A", does not make the hypothetical-conditional statement true by logic of the test.

Scenario II: If "A" is true and the consequent "C" is false, as when the theory conclusively makes erroneous predictions, then the theory is falsified, because the hypothetical conditional "For every A if A, then C" is false. Falsification occurs when the prediction statements in "C" and the observation reports in "O" are not accepted as describing the same thing within the range of vagueness and/or measurement error, which are manifestations of empirical under-determination. The falsifying logic of the test is the *modus tollens* argument form, according to which the conditional-hypothetical schema expressing the theory is falsified, when one affirms the antecedent clause and denies the consequent clause. This is the falsificationist philosophy of scientific criticism advanced by Peirce, the founder of classical pragmatism, and later advocated by Popper.

For more on Popper readers are referred to **BOOK V** at the free web site www.philsci.com or in the e-book *Twentieth-Century Philosophy of Science: A History*, which is available from most Internet booksellers.

The response to a conclusive falsification may or may not be attempts to develop a new theory. Responsible scientists will not deny a falsifying outcome of a test, so long as they accept its test design and test execution. Characterization of falsifying anomalous cases is informative, because it contributes to articulation of a new problem that a new and more empirically adequate theory must solve. Some scientists may, as Kuhn said, simply believe that the anomalous outcome is an unsolved problem for the tested theory without attempting to develop a new theory. But such a response is either an *ipso facto* rejection of the tested theory, a *de facto* rejection of the test design or simply a disengagement from attempts to solve the new problem. And contrary to Kuhn this procrastinating response to anomaly need not imply that the falsified theory has been given institutional status, unless the science itself is institutionally retarded.

For more on Kuhn readers are referred to **BOOK VI** at the free web site www.philsci.com or in the e-book *Twentieth-Century Philosophy of Science: A History,* which is available from most Internet booksellers.

Scenario III: If "A" and "C" are both true, then the hypothetical-conditional schema expressing the tested theory is validly accepted as asserting a causal dependency between the phenomena described by the antecedent and consequent clauses. The nontruth-functional hypothetical-conditional statement does not merely assert a Humean psychological constant conjunction. Causality is an ontological category describing a real dependency, and the causal claim is asserted on the basis of ontological relativity due to the empirical adequacy demonstrated by the nonfalsifying test outcome. Because the nontruth-functional hypothetical-conditional statement is empirical, causality claims are always subject to future testing, falsification, and then revision. This is also true when the conditional represents a mathematical function.

But if the test design is afterwards modified such that it changes the characterization of the subject of the theory, then a previous nonfalsifying test outcome should be reconsidered and the theory should be retested for the new definition of the subject. If the retesting produces a falsifying outcome, then the new information in the modification of the test design has made the terms common to the two test designs equivocal and has contributed parts to alternative meanings. But if the test outcome is not falsification, then the new information is merely new parts added to the meaning of the univocal terms common to the old and new test-design description. Such would be the case for a new and additional way to measure temperature for extreme values that cannot be measured by the old measurement procedure, but which yields the same temperature values within the range of measurement errors, where the alternative procedures produce overlapping results.

On the contemporary pragmatist philosophy a theory that has been tested is no longer theory, once the test outcome is known and the test execution is accepted as correct. If the theory has been falsified, it is merely rejected language unless the falsified theory is still useful for the lesser truth it contains. But if it has been tested with a nonfalsifying test outcome, then it is empirically warranted and thus deemed a scientific law until it is later tested again and falsified. The law is still hypothetical because it is empirical, but it is less hypothetical than it had previously been as a theory proposed for testing. The law may thereafter be used either in an explanation or in a test design for testing some other theory.

For example the elaborate engineering documentation for the Large Hadron Collider at CERN, the *Conseil Européen pour la Recherche Nucléaire,* is based on previously tested science. After installation of the collider is complete and it is known to function successfully, the science in that engineering is not what is tested when the particle accelerator is operated for the microphysical experiments, but rather the employed science is presumed true and contributes to the test design semantics for experiments performed with the accelerator.

4.16 Test Logic Illustrated

Consider the simple heuristic case of Gay-Lussac's law for a fixed amount of gas in an enclosed container as a theory proposed for testing. The container's volume is constant throughout the experimental test, and therefore is not represented by a variable. The theory is **(T'/T)*P = P'**, where the variable **P** means gas pressure, the variable **T** means the gas temperature, and the variables **T'** and **P'** are incremented values for **T** and **P** in a controlled experimental test, where **T' = T ± ΔT,** and **P'** is the predicted outcome that is produced by execution of the test design.

The statement of the theory may be schematized in the nontruth-functional hypothetical-conditional form "For every A if A, then C", where "A" includes **(T'/T)*P**, and "C" states the calculated prediction value of **P',** when temperature is incremented by **ΔT** from **T** to **T'.** The theory is universally quantified, and thus claims to be true for every execution of the experimental test. And for proponents of the theory, who are believers in the theory, the semantics of **T, P, T'** and **P'** are mutually contributing to the semantics of each other, a fact exhibited explicitly in this case, because the equation is monotonic, such that each variable can be expressed as a mathematical function of all the others by simple algebraic transformations.

"A" also includes the universally quantified test-design statements. These statements describe the experimental set up, the procedures for executing the test and initial conditions to be realized for execution of a test. They include description of the equipment used including the container, the heat source, the instrumentation used to measure the magnitudes of heat and pressure, and the units of measurement for the magnitudes involved, namely the pressure units in atmospheres and the temperature units in degrees Kelvin (K°). And they describe the procedure for executing the repeatable experiment. This test-design language is also universally quantified and thus also contributes meaning components to the semantics of the variables **P, T** and **T'** in "A" for all interested scientists who accept the test design.

The procedure for performing the experiment must be executed as described in the test-design language, in order for the test to be valid. The procedure will include firstly measuring and recording the initial values of **T** and **P.** For example let **T** = 200°K and **P** = 1.6 atmospheres. Let the incremented measurement value be recorded as **ΔT** = 200°K, so that the measurement value for **T'** is made to be 400°K. The description of the execution of the procedure and the recorded magnitudes are expressed in particularly quantified test-design language for this particular test execution. The value of **P'** is then calculated.

The test outcome consists of measuring and recording the resulting observed incremented value for pressure. Let this outcome be represented by particularly quantified statement **O** using the same vocabulary as in the test design. But only the universally quantified test-design statements define the semantics of **O,** so that the test is independent of the theory. In this simple experiment one can simply denote the measured value for the resulting observed pressure by the variable **O.** The test execution would also likely be repeated to enable estimation of the range of measurement error in **T**, **T'**, **P** and **O,** and the measurement error propagated into **P'** by calculation. A mean average of the measurement values from repeated executions would be calculated for each of these variables. Deviations from the mean are estimates of the amounts of measurement error, and statistical standard deviations could summarize the dispersion of measurement errors about the mean averages.

The mean average of the test-outcome measurements for **O** is compared to the mean average of the predicted measurements for **P'** to determine the test outcome. If the values of **P'** and **O** are equivalent within their estimated ranges of measurement error, *i.e.*, are sufficiently close to 3.2 atmospheres as to be within the measurement errors, then the theory is deemed not to have been falsified. After repetitions with more extreme incremented values with no falsifying outcome, the theory will likely be deemed sufficiently warranted empirically to be called a law, as it is today.

4.17 Semantics of Empirical Testing

Much has already been said about the artifactual character of semantics, about componential semantics, and about semantical rules. In the semantical discussion that follows, these concepts are brought to bear upon the discussion of the semantics of empirical testing and of test outcomes.

The ordinary semantics of empirical testing is as follows:

If a test has a nonfalsifying outcome, then for the theory's developer and its advocates the semantics of the tested theory is unchanged.

Since they had proposed the theory in the belief that it would not be falsified, their belief in the theory makes it function for them as a set of semantical rules. Thus for them both the theory and the test design are accepted as true, and after the nonfalsifying test outcome both the theory and test-design statements continue to contribute parts to the complex meanings of the descriptive terms common to both theory and test design, as before the test.

But if the test outcome is a falsification, then there is a semantical change produced in the theory for the developer and the advocates of the tested theory who accept the test outcome as a falsification.

The unchallenged test-design statements continue to contribute semantics to the terms common to the theory and test design by contributing their parts – their semantic values – to the meaning complexes of each of those common terms. But the component parts of those meanings contributed by the falsified theory statements are excluded from the semantics of those common terms for the proponents who no longer believe in the theory due to the falsifying test, because the falsified theory statements are no longer deemed to be semantical rules. However, it may be noted incidentally that even if the descriptive terms in the falsifying statement are made to retain the semantic values supplied by the theory, the statement used to describe the test outcome will still have contradicted and thus falsified the theory and in the theory's own terms, i.e., with the theory's semantics.

4.18 Test Design Revision

Empirical tests are conclusive decision procedures only for scientists who agree on which language is proposed theory and which language is presumed test design, and who furthermore accept both the test design and the test-execution outcomes produced with the accepted test design.

The decidability of empirical testing is not absolute. Popper had recognized that the statements reporting the observed test outcome, which he called "basic statements", require agreement by the cognizant scientists, and that those basic statements are subject to future reconsideration.

All universally quantified statements are hypothetical, but theory statements are relatively more hypothetical than test-design statements, because the interested scientists agree that in the event of a falsifying test outcome, revision of the theory will likely be more productive than revision of the test design.

But a dissenting scientist who does not accept a falsifying test outcome of a theory has either rejected the report of the observed test outcome or reconsidered the test design. If he has rejected the outcome of the individual test execution, he has merely questioned whether or not the test was executed in compliance with its agreed test design. Independent repetition of the test with conscientious fidelity to the design may answer such a challenge to the test's validity one way or the other.

But if in response to a falsifying test outcome the dissenting scientist has reconsidered the test design itself, he has thereby changed the semantics involved in the test in a fundamental way. Such reconsideration amounts to rejecting the design as if it was falsified, and letting the theory define the subject of the test and the problem under investigation – **a rôle**

reversal in the pragmatics of test-design language and theory language that makes the original test design and the falsifying test execution irrelevant.

In his "Truth, Rationality, and the Growth of Knowledge" (1961) reprinted in *Conjectures and Refutations* (1963) Popper rejects such a dissenting response to a test, calling it a "content-decreasing stratagem". He admonishes that the fundamental maxim of every critical discussion is that one should "stick to the problem". But as Conant recognized to his dismay in his *On Understanding Science: An Historical Approach* (1947) the history of science is replete with such prejudicial responses to scientific evidence that have nevertheless been productive and strategic to the advancement of basic science in historically important episodes. The prejudicially dissenting scientists may decide that the design for the falsifying test supplied an inadequate description of the problem that the tested theory is intended to solve, often if he developed the theory himself and did not develop the test design. The semantical change produced for such a recalcitrant believer in the theory affects the meanings of the terms common to the theory and test-design statements. The parts of the meaning complex that had been contributed by the rejected test-design statements are parts that are excluded from the semantics of one or several of the descriptive terms common to the theory and test-design statements. Such a semantical outcome can indeed be said to be "content decreasing", as Popper said.

But a scientist's prejudiced or "tenacious" (per Feyerabend) rejection of an apparently falsifying test outcome may have a contributing function in the development of science. It may function as what Feyerabend called a "detecting device", a practice he called **"counterinduction"**, which is a strategy that he illustrated in his examination of Galileo's arguments for the Copernican cosmology. Galileo used the apparently falsified heliocentric theory as a "detecting device" by letting his prejudicial belief in the heliocentric theory control the semantics of the apparently falsifying observational description. This enabled Galileo to reinterpret observations previously described with the equally prejudiced alternative semantics built into the Aristotelian geocentric cosmology. Counterinduction was also the strategy used by Heisenberg, when he reinterpreted the observational description of the electron track in the Wilson cloud chamber using Einstein's aphorism that the theory decides what the physicist can observe, and Heisenberg reports that he then developed his indeterminacy relations using his matrix-mechanics quantum concepts.

Another historic example of using an apparently falsified theory as a detecting device is the discovery of the planet Neptune. In 1821, when Uranus happened to pass Neptune in its orbit – an alignment that had not occurred since 1649 and was not to occur again until 1993 – Alexis Bouvard (1767-1843) developed calculations predicting future positions of the planet Uranus using Newton's celestial mechanics. But observations of Uranus showed significant deviations from the predicted positions.

A <u>first</u> possible response would have been to dismiss the deviations as measurement errors and preserve belief in Newton's celestial mechanics. But astronomical measurements are repeatable, and the deviations were large enough that they were not dismissed as observational errors. The deviations were recognized to have presented a new problem.

A <u>second</u> possible response would have been to give Newton's celestial mechanics the hypothetical status of a theory, to view Newton's law of gravitation as falsified by the anomalous observations of Uranus, and then to attempt to revise Newtonian celestial mechanics. But by then confidence in Newtonian celestial mechanics was very high, and no alternative to Newton's physics had yet been proposed. Therefore there was great reluctance to reject Newtonian physics.

A third possible response, which was historically taken, was to preserve belief in the Newtonian celestial mechanics, to modify the test-design language by proposing a new auxiliary hypothesis of a gravitationally disturbing planet, and then to reinterpret the observations by supplementing the description of the deviations using the auxiliary hypothesis. Disturbing phenomena can "contaminate" even supposedly controlled laboratory experiments. The auxiliary hypothesis changed the semantics of the test-design description with respect to what was observed.

In 1845 both John Couch Adams (1819-1892) in England and Urbain Le Verrier (1811-1877) in France independently using apparently falsified Newtonian physics as a detecting device made calculations of the positions of a disturbing postulated planet to guide future observations in order to detect the postulated disturbing body by telescope. On 23 September 1846 using Le Verrier's calculations Johann Galle (1812-1910) observed the postulated planet with the telescope of the Royal Observatory in Berlin.

Theory is language proposed for testing, and test design is language presumed for testing. But here the pragmatics of the discourses was reversed. In this third response the Newtonian gravitation law was not deemed a tested and falsified theory, but rather was presumed to be true and used for a new test design. The new test-design language was actually given the relatively more hypothetical status of theory by the auxiliary hypothesis of the postulated planet thus newly characterizing the observed deviations in the positions of Uranus. The nonfalsifying test outcome of this new hypothesis was Galle's observational detection of the postulated planet, which Le Verrier had named Neptune.

But counterinduction is after all just a strategy, and it is more an exceptional practice than the routine one. Le Verrier's counterinduction strategy failed to explain a deviant motion of the planet Mercury when its orbit comes closest to the sun, a deviation known as its perihelion precession. In 1843 Le Verrier presumed to postulate a gravitationally disturbing planet that he named Vulcan and predicted its orbital positions. However unlike Le Verrier, Einstein had given Newton's celestial mechanics the more hypothetical status of theory language, and he viewed Newton's law of gravitation as having been falsified by the anomalous perihelion precession. He had initially attempted a revision of Newtonian celestial mechanics by generalizing on his special theory of relativity. This first such attempt is known as his *Entwurf* version, which he developed in 1913 in collaboration with his mathematician friend Marcel Grossman. But working in collaboration with his friend Michele Besso he found that the *Entwurf* version had clearly failed to account accurately for Mercury's orbital deviations; it showed only 18 seconds of arc per century instead of the actual 43 seconds.

In 1915 he finally abandoned the *Entwurf* version, and under prodding from the mathematician David Hilbert (1862-1943) he turned to mathematics exclusively to produce his general theory of relativity. He then developed his general theory, and announced his correct prediction of the deviations in Mercury's orbit to the Prussian Academy of Sciences on 18 November 1915. He received a congratulating letter from Hilbert on "conquering" the perihelion motion of Mercury. After years of delay due to World War I his general theory was further vindicated by Arthur Eddington's (1888-1944) historic eclipse test of 1919. Some astronomers reported that they had observed a transit of a planet across the sun's disk, but these claims were found to be spurious when larger telescopes were used, and Le Verrier's postulated planet Vulcan has never been observed. MIT professor Thomas Levenson relates the history of the futile search for Vulcan in his *The Hunt for Vulcan* (2015).

Le Verrier's response to Uranus' deviant orbital observations was the opposite to Einstein's response to the deviant orbital observations of Mercury. Le Verrier reversed the rôles of theory and test-design language by preserving his belief in Newton's physics and using it to revise the test-design language with his postulate of a disturbing planet. Einstein viewed Newton's celestial mechanics to be hypothetical, because he believed that the Newtonian theory statements were more likely to be productively revised than test-design statements, and he took the anomalous orbital observations of Mercury to falsify Newton's physics, thus indicating that theory revision was needed. Empirical tests are conclusive decision procedures only for scientists who agree on which language is proposed theory and which is presumed test design, and who furthermore accept both the test design and the test-execution outcomes produced with the accepted test design.

For more about Feyerabend on counterinduction readers are referred to **BOOK VI** at the free web site www.philsci.com or in the e-book *Twentieth-Century Philosophy of Science: A History*, which is available from most Internet booksellers.

Finally there are more routine cases of test design revision that do not occasion counterinduction. In such cases there is no rôle reversal in the pragmatics of theory and test design, but there may be an equivocating revision in the test-design semantics depending on the test outcome due to a new observational technique or instrumentality, which may have originated in what Feyerabend called "auxiliary sciences", *e.g.*, development of a superior microscope or telescope. If retesting a previously nonfalsified theory with the new test design with the new observational technique or instrumentality does not produce a falsifying outcome, then the result is merely a refinement of the semantics in the test-design language. But if the new test design occasions a falsification, then it has produced a semantical equivocation between the statements of the old and new test-designs, and has redefined the subject of the tested theory.

4.19 Empirical Underdetermination

Conceptual vagueness and measurement error are manifestations of empirical underdetermination, which may occasion scientific pluralism.

The empirical underdetermination of language may make empirical criteria incapable of producing a decisive theory-testing outcome. Two manifestations of empirical underdetermination are conceptual vagueness and measurement error. All concepts have vagueness that can be reduced indefinitely but can never be eliminated completely. Mathematically expressed theories use measurement data that always contain measurement inaccuracy that can be reduced indefinitely but never eliminated completely.

Scientists prefer measurements and mathematically expressed theories, because they can measure the amount of prediction error in the theory, when the theory is tested. But separating measurement error from a theory's prediction error can be problematic. Repeated careful execution of the measurement procedure, if the test is repeatable, enables statistical estimation of the range of measurement error. But in research using historical time-series data such as economics, repetition is impossible.

4.20 Scientific Pluralism

Scientific pluralism is recognition of the coexistence of multiple empirically adequate alternative explanations due to undecidability resulting from the empirical underdetermination in a test-design.

All language is always empirically underdetermined by reality. Empirical underdetermination explains how two or more semantically alternative empirically adequate explanations can have the same test-design. This means that there are several theories having alternative explanatory factors and yielding accurate predictions that are alternatives to one another, while predicting differences that are small enough to be within the range of the estimated measurement error in the test design. In such cases empirical underdetermination due to the current test design imposes undecidability on the choice among the alternative explanations.

Econometricians are accustomed to alternative empirically adequate econometric models. This occurs because measurement errors in aggregate social statistics are typically large in comparison to those available in laboratory sciences. Each such model has different equation specifications, i.e., different causal variables in the equations of the model, and makes different predictions for some of the same prediction variables that are accurate within the relatively large range of estimated measurement error. And discovery systems with empirical test procedures routinely proliferate empirically adequate alternative explanations for output. They produce what Einstein called "an embarrassment of riches". Logically this multiplicity of alternative explanations means that there may be alternative empirically warranted nontruth-functional hypothetical conditional schemas in the form "For all A if A, then C" having alternative causal antecedents "A" and making different but empirically adequate predictions that are the empirically indistinguishable consequents "C".

Empirical underdetermination is also manifested as conceptual vagueness. For example to develop his three laws of planetary motion Johannes Kepler (1591-1630), a heliocentrist, used the measurement observations of Mars that had been collected by Tycho Brahe (1546-1601), a type of geocentrist. Brahe had an awkward geocentric-heliocentric cosmology, in which the fixed earth is the center of the universe, the stars and the sun revolve around the earth, and the other planets revolve around the sun. Kepler used Brahe's astronomical measurement data. There was empirical underdetermination in these measurement data, as in all measurement data.

Measurement error was not the operative empirical underdetermination permitting the alternative cosmologies, because both used the same data. But Kepler was a convinced Copernican placing the sun at the center of the universe. His belief in the Copernican heliocentric cosmology made the semantic parts contributed by that heliocentric cosmology become for him component parts of the semantics of the language used for celestial observation, thus displacing Brahe's more complicated combined geocentric-heliocentric cosmology's semantical contribution. The manner in which Brahe and Kepler could have different observations is discussed by Hanson in his chapter "Observation" in his *Patterns of Discovery*. Hanson states that even if both astronomers saw the same dawn, they nonetheless saw differently, because observation depends on the conceptual organization in one's prior knowledge and language. Hanson uses the "see that…." locution. Thus Brahe *sees that* the sun is beginning its journey from horizon to horizon, while Kepler *sees that* the earth's horizon is dipping away from our fixed local star. Einstein said that the theory decides what the physicist can observe; Hanson similarly said that observation is "theory laden".

Alternative empirically adequate explanations due to empirical underdetermination are all more or less true. An answer as to which explanation is truer must await further development of additional observational information or measurements that reduce the empirical underdetermination in the test-design concepts. But there is never any ideal test design with "complete" information, *i.e.,* without vagueness or measurement error. Recognition of possible undecidability among alternative empirically adequate scientific explanations due to empirical underdetermination occasions what pragmatists call "**scientific pluralism**".

4.21 Scientific Truth

Truth and falsehood are spectrum properties of statements, such that the greater the truth, the lesser the error.

Tested and nonfalsified statements are more empirically adequate, have more realistic ontologies, and are truer than tested and falsified statements.

Falsified statements have recognized error, and may simply be rejected, unless they are still useful for their lesser realism and truth.

What is truth! Truth is a spectrum property of descriptive language with its relativized semantics and ontology. It is not merely a subjective expression of approval.

Belief and truth are not identical. Belief is acceptance of a statement as predominantly true. As Jarrett Leplin maintains in his *Defense of Scientific Realism* (1997), truth and falsehood are properties that admit to more or less. Thus they are not simply dichotomous, as they are represented in two-valued formal logic. Therefore one may wrongly believe that a predominantly false statement is predominantly true, or wrongly believe that a predominantly true statement is predominantly false. Belief controls the semantics of the descriptive terms in a universally quantified statement, while truth is the relation of a statement's semantics and the ontology it describes to mind-independent nonlinguistic reality.

Test-design language is presumed true with definitional force for the semantics of the test-design language, in order to characterize the subject and procedures of a test. Theory language in an empirical test may be believed true by the developer and advocates of the theory, but the theory is not true simply by virtue of their belief. Belief in an untested theory is speculation about a future test outcome. A nonfalsifying test outcome will warrant belief that the tested theory is as true as the theory's demonstrated empirical adequacy. Empirically falsified theories have recognized error, are predominantly false, and may be rejected unless they are still useful for their lesser realism and lesser truth. Tested and nonfalsified statements are more empirically adequate, have ontologies that are more realistic, and thus are truer than empirically falsified statements.

Popper said that Eddington's historic eclipse test of Einstein's theory of gravitation in 1919 "falsified" Newton's theory and thus "corroborated" Einstein's theory. Yet the U.S. National Aeronautics and Space Administration (NASA) today still uses Newton's laws to navigate interplanetary rocket flights such as the *Voyager* and *New Horizon* missions. Thus Newton's "falsified" theory is not completely false or totally unrealistic, or it could never have been used before or after Einstein. Popper said that science does not attain truth. But contemporary pragmatists believe that such an absolutist idea of truth is misconceived. Advancement in empirical adequacy is advancement in realism and in truth. Feyerabend said, “Anything goes”. Regarding ontology Hickey says, “Everything goes”, because while not all discourses are equally valid, there is no semantically interpreted syntax utterly devoid of ontological significance and thus no discourse utterly devoid truth. Therefore Hickey adds that the more empirically adequate explanation goes farther – is truer and more realistic – than its less empirically adequate falsified alternatives.

Empirical science progresses in empirical adequacy, and thereby in realism and in truth.

4.22 Nonempirical Criteria

Confronted with unresolvable scientific pluralism – having several alternative explanations that are tested and not falsified due to empirical underdetermination in the test-design language – philosophers and scientists have proposed various nonempirical criteria that they believe have been operative historically in explanation choice.

And a plurality of untested and therefore unfalsified theories may also exist before any testing, so that different scientists may have their preferences for testing one theory over another based on nonempirical criteria.

Philosophers have proposed a variety of such nonempirical criteria. Popper advances a criterion that he says enables the scientist to know in advance of any empirical test, whether or not a new theory would be an improvement over existing theories, were the new theory able to pass crucial tests, in which its performance is comparable to older existing alternatives. He calls this criterion the “potential satisfactoriness” of the theory, and it is measured by the amount of “information content” in the theory. This criterion follows from his concept of the aim of science, the thesis that the theory that tells us more is preferable to one that tells us less, because the more informative theory has more "potential falsifiers".

But a theory with greater potential satisfactoriness may be empirically inferior, when tested with an improved test design. Test designs are improved by developing more accurate measurement procedures and/or by adding new descriptive information that reduces the vagueness in the characterization of the subject for testing. Such test-design improvements refine the characterization of the problem addressed by the theories, and thus reduce empirical underdetermination to improve the decidability of testing.

When empirical underdetermination makes testing undecidable among alternative theories, different scientists may have personal reasons for preferring one or another alternative as an explanation. In such circumstances selection may be an investment decision for the career scientist rather than an investigative decision. The choice may be influenced by such circumstances as the cynical realpolitik of peer-reviewed journals. Knowing what editors and their favorite referees currently want in submissions helps an author getting his paper published. Publication is an academic status symbol with the more prestigious journals yielding more brownie points for accumulating academic tenure, salary and status.

In the January 1978 issue of the *Journal of the American Society of Information Science* (JASIS) the editor wrote that referees often use the peer review process as a means to attack a point of view and to suppress the content of a submitted paper, *i.e.*, they attempt censorship. Furthermore editors are not typically entrepreneurial; as "gate guards" they are academia's risk-aversive rearguard rather than the risk-taking avant-garde. They select the established "authorities" with reputation-based vested interests in the prevailing traditional views. These so-called authorities cynically suborn the peer-review process by using their conventional views as criteria for criticism and rejection for publication instead of empirical criteria. Such cynical reviewers and editors are effectively hacks that represent the *status quo* demanding trite papers rather than new, original and empirically superior ideas. When this conventionality producing hacks becomes sufficiently pervasive, it becomes normative, which is to say the science has become institutionally corrupted. In contemporary academic sociology conventionality is accentuated by the conformism that is highly valued by sociological theory, which among sociologists is still furthermore reinforced by sociologists' enthusiastic embrace of Kuhn's conformist sociological thesis of "normal science". Corruption has thus become more institutionalized and thus more retarding to development in academic sociology than in the other sciences.

External sociocultural factors have also influenced theory choice. In his *Copernican Revolution: Planetary Astronomy in the Development of Western Thought* (1957) Kuhn wrote that the astronomer in the time of Copernicus could not upset the two-sphere universe without overturning physics and religion as well. He reports that fundamental concepts in the pre-Copernican astronomy had become strands for a much larger fabric of thought, and the nonastronomical strands in turn bound the thinking of the astronomers. The Copernican revolution occurred because Copernicus was a dedicated specialist, who valued mathematical and celestial detail more than the values reinforced by the nonastronomical views that were dependent on the prevailing two-sphere theory. This purely technical focus of Copernicus enabled him to ignore the nonastronomical consequences of his innovation, consequences that would lead his contemporaries of less restricted vision to reject his innovation as absurd.

Later in discussing modern science in his popular *Structure of Scientific Revolutions* Kuhn does not make the consequences for the nonspecialist an aspect of his general theory of scientific revolutions. Instead he maintains, as part of his thesis of "normal" science that a scientist may willfully choose to ignore a falsifying outcome of a decisive test execution. This choice is not due to the scientist's effective criticism of either the test design or the test execution, but rather is due to the expectation that the falsified theory will later be improved and corrected. However any such "correcting" alteration made a to falsified theory amounts to a discovery strategy that Hickey calls "theory elaboration" that produces a new and different theory (See above, Section **4.12**).

Citing Kuhn some sociologists of knowledge including those advocating the "strong program" maintain that the social and political forces that influence society at large also inevitably influence the content of scientific beliefs. This is truer in the social sciences, but sociologists who believe that this means empiricism does not control acceptance of scientific beliefs in the long term are mistaken, because it is pragmatic empiricism that enables wartime victories, peacetime prosperity – and in all times business profits – as reactionary politics, delusional ideologies and utopian fantasies cannot.

Persons with different economic views defend and attack certain social/political philosophies, ideologies, special interests and provincial policies. For example in the United States more than eighty years after Keynes, Republican politicians still attack Keynesian economics while Democrat politicians defend it. Many Republicans are motivated by the right-wing political ideology such as may be found in Hayek's *Road to Serfdom* or in the novels by Ayn Rand. The prevailing political philosophy among Republicans opposes government intervention in the private economy. But as Federal Reserve

Board of Governors Chairman Ben Bernanke, New York Federal Reserve Bank President Timothy Geithner and U.S. Treasury Secretary Henry Paulson maintain in their *Firefighting: The Financial Crisis and Its Lessons* (2019), Adam Smith's invisible hand of capitalism cannot stop a full blown financial collapse; only the visible hand of government can do that (P. 5).

Thus pragmatism prevailed over ideology, when expediency dictated, as happened during the 2007-2009 Great Recession crisis. In his *After the Music Stopped* (2013) Alan S. Blinder, Princeton University economist and former Vice Chairman of the Federal Reserve Board of Governors, reports that "ultraconservative" Republican President George W. Bush "let pragmatism trump ideology" (P. 213), when he signed the Economic Stimulus Act of 2008, a distinctively Keynesian fiscal policy of tax cuts, which added $150 billion to the U.S. Federal debt notwithstanding Republicans' visceral abhorrence of the Federal debt.

In contrast Democrat President Barak Obama without reluctance and with a Democrat-controlled Congress signed the American Reinvestment and Recovery Act in 2009, a stimulus package that added $787 billion to the Federal debt. Blinder reports that simulations with the Moody Analytics large macroeconometric model showed that the effect of Obama's stimulus in contrast to a no-stimulus simulation scenario was a GDP that was 6 per cent higher with the stimulus than without it, an unemployment rate 3 percentage points lower, and 4.8 million additional Americans employed (P. 209).

Nonetheless as former Federal Reserve Board Chairman Ben Bernanke wrote in his memoir *The Courage to Act* (2013), the 2009 stimulus was small in comparison with its objective of helping to arrest the deepest recession in seventy years in a $15 trillion national economy (P. 388). Thus

Bernanke, a conservative Republican, did not reject Keynesianism, but instead actually concluded that the recovery was needlessly slow, because the Obama Federal fiscal stimulus program was disproportionately small for the U.S. national macroeconomy.

All nonempirical criteria are presumptuous. **No nonempirical criterion enables a scientist to predict reliably which among alternative nonfalsified explanations will survive empirical testing, when in due course the degree of empirical underdetermination is reduced by a new and improved test design that enables decidable testing.**

To make such anticipatory choices is like betting on a horse before it runs the race.

4.23 The "Best Explanation" Criteria

As previously noted (See above, Section **4.05**) Thagard's cognitive-psychology system **ECHO** developed specifically for theory selection has identified three nonempirical criteria to maximize achievement of the coherence aim. His simulations of past episodes in the history of science indicate that the most important criterion is **breadth of explanation**, followed by **simplicity of explanation**, and finally **analogy with previously accepted theories.** Thagard considers these nonempirical selection criteria as productive of a "best explanation".

The breadth-of-explanation criterion also suggests Popper's aim of maximizing information content. In any case there have been successful theories in the history of science, such as Heisenberg's matrix mechanics and uncertainty relations, for which none of these three characteristics were operative in the acceptance as explanations. And as Feyerabend noted in *Against Method* in criticizing Popper's view,

Aristotelian dynamics is a general theory of change comprising locomotion, qualitative change, generation and corruption, while Galileo and his successors' dynamics pertains exclusively to locomotion. Aristotle's explanations therefore may be said to have greater breadth, but his physics is now deemed to be less empirically adequate.

Contemporary pragmatists acknowledge *only* the empirical criterion, the criterion of superior empirical adequacy. They exclude all nonempirical criteria from the aim of science, because while relevant to persuasion to make theories appear "convincing", they are irrelevant as evidence of progress. Nonempirical criteria are like the psychological criteria that trial lawyers use to select and persuade juries in order to win lawsuits in a court of law, but which are irrelevant to courtroom evidence rules for determining the facts of a case. Such prosecutorial lawyers are like the editors and referees of the peer-reviewed academic literature (sometimes called the "court of science") who ignore the empirical evidence described in a paper submitted for publication and who reject the paper due to its unconventionality. Such editors make marketing-based promotional decisions instead of evidence-based publication decisions.

But nonempirical criteria are often operative in the selection of problems to be addressed and explained. For example the American Economic Association's *Index of Economic Journals* indicates that in the years of the Great Depression the number of journal articles concerning the trade cycle fluctuated in close correlation with the national average unemployment rate with a lag of two years.

4.24 Nonempirical Linguistic Constraints

The constraint imposed upon theorizing by empirical test outcomes is the empirical constraint, the

criterion of superior empirical adequacy. It is the regulating institutionalized cultural value definitive of modern empirical science that is not viewed as an obstacle to be overcome, but rather as a condition to be respected for the advancement of science.

But there are other kinds of constraints that are nonempirical and are retarding impediments that must be overcome for the advancement of science, and they are internal to science in the sense that they are inherent in the nature of language. They are the **cognition constraint** and **communication constraint.**

4.25 Cognition Constraint

The semantics of every descriptive term is determined by its linguistic context consisting of universally quantified statements believed to be true.

Conversely given the conventional meaning for a descriptive term, certain beliefs determining the meaning of the term are reinforced by habitual linguistic fluency with the result that the meaning's conventionality constrains change in those defining beliefs.

The conventionalized meanings for descriptive terms produce the cognition constraint. The cognition constraint is the linguistic impediment that inhibits construction of new theories, and is manifested as lack of imagination, creativity or ingenuity.

In his *Concept of the Positron* Hanson identified this impediment to discovery and called it the "conceptual constraint". He reports that physicists' identification of the concept of the subatomic particle with the concept of its charge was an impediment to recognizing the positron. The electron was identified with a negative charge and the much more massive proton was identified with a positive charge, so that the positron as a particle with the mass of an electron and a positive charge was not recognized without difficulty and delay.

In his *Introduction to Metascience* Hickey referred to this conceptual constraint as the "cognition constraint". The cognition constraint inhibits construction of new theories, and is manifested as lack of imagination, creativity or ingenuity. Semantical rules are not just explicit rules; they are also strong linguistic habits with subconscious roots that enable prereflective competence and fluency in both thought and speech. Six-year-old children need not reference explicit grammatical and semantical rules in order to speak competently and fluently. And these subconscious habits make meaning a synthetic psychological experience.

Given a conventionalized belief or firm conviction expressible as a universally quantified affirmative statement, the predicate in that affirmation contributes meaning parts to the meaning complex of the statement's subject term. Not only does the conventionalized status of meanings make development of new theories difficult, but also any new theory construction requires greater or lesser semantical dissolution and restructuring. Accordingly the more extensive the revision of beliefs, the more constraining are both the semantical restructuring and the psychological conditioning on the creativity of the scientist who would develop a new theory. Revolutionary theory development requires both relatively more extensive semantical dissolution and restructuring and thus greater psychological adjustment in linguistic habits.

However, use of computerized discovery systems circumvents the cognition constraint, because the machines have no linguistic-psychological habits and make no semantical interpretations. Their mindless electronic execution of mechanized procedures is one of their virtues.

The cognition-constraint thesis is opposed to the neutral-language thesis that language is merely a passive instrument for expressing thought. Language is not merely passive but rather has a formative influence on thought. The formative influence of language as the "shaper of meaning" has been recognized as the Sapir-Whorf hypothesis and specifically by Benjamin Lee Whorf's principle of linguistic relativity set forth in his "Science and Linguistics" (1940) reprinted in *Language, Thought and Reality* (1956). But contrary to Whorf it is not the grammatical system that determines semantics, but rather it is what Quine called the "web of belief", *i.e.*, the shared belief system as found in a unilingual dictionary.

For more about the linguistic theory of Whorf readers are referred to in **BOOK VI** at the free web site www.philsci.com or in the e-book *Twentieth-Century Philosophy of Science: A History*, which is available from most Internet booksellers.

4.26 Communication Constraint

The communication constraint is the linguistic impediment to understanding a new theory relative to those currently conventional.

The communication constraint has the same origins as the cognition constraint. This impediment is also both cognitive and psychological. The scientist must cognitively learn the new theory well enough to restructure the composite meaning complexes associated with the descriptive terms common both to the old theory that he is familiar with and to the theory that is new to him. And this learning involves overcoming psychological habit that enables linguistic fluency that reinforces existing beliefs.

This learning process suggests the conversion experience described by Kuhn in revolutionary transitional episodes, because the new theory must firstly be accepted as true however provisionally for its semantics to be understood, since only statements believed to be true can operate as semantical rules that convey understanding. That is why dictionaries are presumed not to contain falsehoods. If testing demonstrates the new theory's superior empirical adequacy, then the new theory's pragmatic acceptance should eventually make it the established conventional wisdom.

But if the differences between the old and new theories are so great as perhaps to be called revolutionary, then some members of the affected scientific profession may not accomplish the required learning adjustment. People usually prefer to live in an orderly world, but innovation creates semantic disorientation and consequent psychological anomie. In reaction the slow learners and nonlearners become a rearguard that clings to the received conventional wisdom, which is being challenged by the new theory at the frontier of research, where there is much conflict that produces confusion due to semantic dissolution and consequent restructuring of the relevant concepts in the web of belief.

The communication constraint and its effects on scientists have been insightfully described by Heisenberg, who personally witnessed the phenomenon when his quantum theory was firstly advanced. In his *Physics and Philosophy*: *The Revolution in Modern Science* Heisenberg defines a "revolution" in science as a change in thought pattern, which is to say a semantical change, and he states that a change in thought pattern becomes apparent, when words acquire meanings that are different from those they had formerly. The central question that Heisenberg brings to the phenomenon of revolution in science understood as a change in thought pattern is how the revolution is able to come about. The occurrence of a scientific revolution is problematic due to resistance to the change in thought pattern presented to the cognizant profession.

Heisenberg notes that as a rule the progress of science proceeds without much resistance or dispute, because the scientist has by training been put in readiness to fill his mind with new ideas. But he says the case is altered when new phenomena compel changes in the pattern of thought. Here even the most eminent of physicists find immense difficulties, because a demand for change in thought pattern may create the perception that the ground has been pulled from under one's feet. He says that a researcher having achieved great success in his science with a pattern of thinking he has accepted from his young days, cannot be ready to change this pattern simply on the basis of a few novel experiments. Heisenberg states that once one has observed the desperation with which clever and conciliatory men of science react to the demand for a change in the pattern of thought, one can only be amazed that such revolutions in science have actually been possible at all. It might be added that since the prevailing conventional view has usually had time to be developed into a more extensive system of ideas, those unable to cope with the semantic dissolution produced by the newly emergent ideas often take refuge in the psychological comforts of coherence and familiarity provided by the more extensive conventional wisdom, which assumes the nature of a dogma and for some scientists an ideology.

In the meanwhile the developers of the new ideas together with the more opportunistic and typically younger advocates of the new theory, who have been motivated to master the new theory's language in order to exploit its perceived career promise, assume the avant-garde rôle and become a vanguard. 1970 Nobel-laureate economist Paul Samuelson offers a documented example: He wrote in "Lord Keynes and the General Theory" in *Econometrica* (1946) that he considers it a priceless advantage to have been an economist before 1936, the publication year of Keynes' *General Theory*, and to have received a thorough grounding in classical economics, because his rebellion against Keynes' *General Theory's* pretensions would have been complete save for his uneasy realization that he did not at all understand what it is about. And he adds that no one else in Cambridge, Massachusetts really knew what it is about for some twelve to eighteen months after its publication. Years later he wrote in his *Keynes' General Theory*: *Reports of Three Decades* (1964) that Keynes' theory had caught most economists under the age of thirty-five with the unexpected virulence of a disease first attacking and then decimating an isolated tribe of South Sea islanders, while older economists were the rearguard that was immune. Samuelson was a member of the Keynesian vanguard.

Note also that contrary to Kuhn and especially to Feyerabend the transition however great does not involve a complete semantic discontinuity much less any semantic incommensurability. And it is unnecessary to learn the new theory as though it were a completely foreign language. The semantic incommensurability muddle is resolved by recognition of componential semantics. For the terms common to the new and old theories, the component parts contributed by the new theory replace those from the old theory, while the parts contributed by the test-design statements remain unaffected. Thus the test-design language component parts shared by both theories enable characterization of the subject of both theories independently of the distinctive claims of either, and thereby enable decisive testing. The shared semantics in the test-design language also facilitates learning and understanding the new theory, however radical the new theory may be.

It may furthermore be noted that the scientist viewing the computerized discovery system output experiences the same communication impediment with the machine output that he would, were the outputted theories developed by a fellow human scientist. The communication constraint makes new theories developed mechanically grist for Luddites' mindless rejection.

Fortunately today the Internet and e-book media enable new ideas to circumvent obstructionism by the peer-review literature, functioning as a latter day *Salon des Refusés* for both scientists and philosophers of science. Hickey's communications with sociology journal editors exemplify the retarding effects of the communication constraint in current academic sociology. See **Appendix II** in **BOOK VIII** at the free web site www.philsci.com or in the e-book *Twentieth-Century Philosophy of Science*: *A History*, which is available from most Internet booksellers.

The communication constraint is a general linguistic phenomenon that is not limited to the language of science. It applies to philosophy as well. Many philosophers of science who received much if not all of their philosophy education before the turbulent 1960's or whose philosophy education was for whatever reason retarded, are unsympathetic to the reconceptualization of familiar terms such as "theory" and "law" that are central to contemporary pragmatism. They are dismayed by the semantic dissolution resulting from the rejection of the positivist or romantic beliefs.

In summary both the cognition constraint and the communication constraint are based on the reciprocal relation between semantics and belief, such that given the conventionalized meaning for a descriptive term, certain beliefs determine the meaning of the term, which beliefs are furthermore reinforced by psychological habit that enables linguistic fluency. The result is that the meaning's conventionality impedes change in those defining beliefs.

4.27 Scientific Explanation

The logic for the rational reconstruction of a scientific explanation is as follows:

A scientific explanation is:

(1) a discourse that can be schematized as a *modus ponens* logical deduction from a set of one or several universally quantified law statements expressible in a nontruth-functional hypothetical-conditional schema

(2) together with a particularly quantified antecedent description of realized initial conditions

(3) that jointly conclude to a consequent particularly quantified description of the explained event.

Explanation is the ultimate aim of basic science. There are nonscientific types such as the historical explanation, but history is not a science, although it may use science as in economic history. But only explanation in basic science is of interest in philosophy of science. When some course of action is taken in response to an explanation such as a social policy, a medical therapy or an engineered product or structure, the explanation is used as applied science. Applied science does not occasion a change in an explanation as in basic science, unless there is an unexpected failure in spite of conscientious and competent implementation of the relevant applied laws.

Since a theory in an empirical test is proposed as an explanation, the logical form of the explanation in basic science is the same as that of the empirical test. The universally quantified statements constituting a system of one or several related scientific laws in an explanation can be schematized as a nontruth-functional conditional statement in the logical form "For every A if A, then C". But while the logical form is the same for both testing and explanation, the deductive argument is not the same.

The deductive argument of the explanation is the *modus ponens* argument instead of the *modus tollens* logic used for testing. In the *modus tollens* argument the conditional statement expressing the proposed theory is falsified, when the antecedent clause is true and the consequent clause is false. On the other hand in the *modus ponens* argument for explanation both the antecedent clause describing initial and exogenous conditions and the conditional statements having law status are accepted as true, such that affirmation of the antecedent clause validly concludes to affirmation of the consequent clause describing the explained phenomenon.

Thus the schematic form of an explanation is "For every A if A, then C" is true. "A" is true. Therefore "C" is true (and explained). The conditional statement "For every A if A, then C" represents a set of one or several related universally quantified law statements applying to all instances of "A". "A" is the set of one or several particularly quantified statements describing the realized initial and exogenous conditions that cause the occurrence of the explained phenomenon as in a test. "C" is the set of one or several particularly quantified statements describing the explained individual consequent effect, which whenever possible is a prediction.

In the scientific explanation the statements in the conditional schema express scientific laws accepted as true due to their empirical adequacy as demonstrated by nonfalsifying test outcomes. These together with the antecedent statements describing the initial conditions in the explanation constitute the explaining language that Popper calls the "*explicans*". And he calls the logically consequent language, which describes the explained phenomenon, the "*explicandum*". Hempel used the terms "*explanans*" and "*explanandum*" respectively. Furthermore it has been said that theories "explain" laws. Neither untested nor falsified theories occur in a scientific explanation. Scientific explanations consist of laws, which are formerly theories that have been tested with nonfalsifying test outcomes. Proposed explanations are merely untested theories.

Since all the universally quantified statements in the nontruth-functional conditional schema of an explanation are laws, the "explaining" of laws is said to mean that a system of logically related laws forms a deductive system partitioned into dichotomous subsets of explaining antecedent axioms and explained consequent theorems. Logically integrating laws into axiomatic systems confers psychological satisfaction by contributing semantical coherence. Influenced by Newton's physics many positivists had believed that producing reductionist axiomatic systems is part of the aim of science. Logical reductionism was integral to the positivist Vienna Circle's unity-of-science agenda. Hanson calls this "catalogue science" as opposed to "research science". The logical axiomatizing reductionist fascination is not validated by the history of science. Great developmental episodes in the history of science such as the development of quantum physics have had the opposite effect, *i.e.,* that of fragmenting science, because quantum mechanics cannot be made a logical extension of classical physics. But while fragmentation has occasioned the communication constraint and thus provoked opposition to a discovery, it has delayed but not halted the empirical advancement of science in its history. The ***only*** criterion for scientific criticism that is acknowledged by the contemporary pragmatist is the *empirical criterion.* Eventually empirical pragmatism prevails.

However, physical reductionism as opposed to mere axiomatic logical reductionism represents discoveries in science and does more than just add semantical coherence. Simon and his associates developed discovery systems that produced physical reductions in chemistry. Three such systems named, **STAHL DALTON** and **GLAUBER** are described in Simon's *Scientific Discovery*. System **STAHL,** named after the German chemist Georg Ernst Stahl was developed by Jan Zytkow. It creates a type of qualitative law that Simon calls "componential", because it describes the hidden components of substances. **STAHL** replicated the development of both the phlogiston and the oxygen theories of combustion. System **DALTON,** named after John Dalton the chemist creates structural laws in contrast to **STAHL,** which creates componential laws. Like the historical Dalton the **DALTON** system does not invent the atomic theory of matter. It employs a representation that embodies the hypothesis and incorporates the distinction between atoms and molecules invented earlier by Amadeo Avogado.

System **GLAUBER** was developed by Pat Langley in 1983. It is named after the eighteenth century chemist Johann Rudolph Glauber who contributed to the development of the acid-base theory. Note that the componential description does not invalidate the higher-order description. Thus the housewife who combines baking soda and vinegar and then observes a reaction yielding a salt residue may validly and realistically describe the vinegar and soda (acid and base) and their observed reaction in the colloquial terms she uses in her kitchen. The colloquial description is not invalidated by her inability to describe the reaction in terms of the chemical theory of acids and bases. Both descriptions are semantically significant and each realistically describes an ontology.

The difference between logical and physical reductions is illustrated by the neopositivist Ernest Nagel in his distinction between "homogeneous" and "heterogeneous" reductions in his *Structure of Science* (1961). The homogeneous reduction illustrates what Hanson called "catalogue science", which is merely a logical reduction that contributes semantical coherence, while the heterogeneous reduction illustrates what Hanson called "research science", which involves discovery and new empirical statements that Nagel calls "correspondence rules". In the case of the homogeneous reduction, which is merely a logical reduction with one of the theories operating as a set of axioms and the other as a set of conclusions, the semantical effect is merely an exhibition of semantical structure and a decrease in vagueness to increase coherence. This can be illustrated by the reduction of Kepler's laws describing the orbital motions of the planet Mars to Newton's laws of gravitation.

However in the case of the heterogeneous reduction there is not only a reduction of vagueness, but also the addition of correspondence rules, which are universally quantified falsifiable empirical statements relating descriptive terms in the two theories to one another. Nagel maintains that the correspondence rules are initially conventions that merely assign additional meaning, but which later become testable and falsifiable empirical statements. Nagel illustrates this heterogeneous type by the reduction of thermodynamics to statistical mechanics, in which a temperature measurement value is equated to a measured value of the mean of molecular kinetic energy by a correspondence rule. Then further development of the theory makes it possible to calculate the temperature of the gas in some indirect fashion from experimental data other than the temperature value obtained by actually measuring the temperature of the gas. Thus the molecular kinetic energy laws empirically explain the thermodynamic laws.

In his "Explanation, Reduction and Empiricism" in *Minnesota Studies in the Philosophy of Science* (1962) Feyerabend with his wholistic view of the semantics of language dismissed Nagel's positivist analysis of reductionism. Feyerabend maintained that the reduction is actually a complete replacement of one theory together with its observational consequences with another theory with its distinctive observational consequences. But the contemporary pragmatist can analyze the language of reductions by means of the componential semantics thesis applied to both theories and their shared test design language.

Bibliography

Achinstein, Peter and Stephen F. Baker. (Ed.) *The Legacy of Logical Positivism.* John Hopkins Press, Baltimore, MD, 1969.

Ackley, Gardner. *Macroeconomics: Theory and Policy.* Macmillan Publishing Co., Inc., NY, 1978, [1961]. Pp. 376-382.

Alston, William P. *Philosophy of Language*. Prentice-Hall, Englewood Cliffs, NJ, 1964.

American Economic Association. *Readings in Business Cycles.* Edited by Robert A. Gordon and Lawrence Klein. Richard D. Irwin, Homewood, IL, 1965.

Amirizadeh, Hossain, and Richard M. Todd. "More Growth Ahead for the Ninth District States," *Quarterly Review*, Federal Reserve Bank of Minneapolis (Fall 1984).

Anderson, Fulton H. *Philosophy of Francis Bacon.* Octagon Books, New York, NY, 1971.

Arbib, Michael A. and Mary B. Hesse. *The Construction of Reality.* Cambridge University Press, Cambridge, England, 1986.

August, Eugene R. *John Stuart Mill, A Mind at Large.* Charles Scribner's & Sons, New York, NY, 1975.

Baggott, James E. *Beyond Measure: Modern Physics, Philosophy, and the Meaning of Quantum Theory.* Oxford University Press, New York, NY, 2003.

Bar-Hillel, Yehoshua and Rudolf Carnap. "Semantic Information", *British Journal for the Philosophy of Science,* Vol. 4 (1953), Pp. 147-157.

Bar-Hillel, Yehoshua. *Language and Information.* Addison-Wesley, Reading, MA, 1964.

Belkin, Nicholas J. "Some Soviet Concepts of Information for Information Science," *American Society for Information Science Journal*. Vol. 26. (January 1975), Pp. 56-60.

Bell, J.S. *Speakable and Unspeakable in Quantum Mechanics: Collected Papers on Quantum Philosophy*. Cambridge University Press, Cambridge, England, 1987.

Beller, Mara. "Bohm and the 'Inevitability' of Acausality" in *Bohmian Mechanics and Quantum Theory: An Appraisal.* Eds. James T. Cushing, Arthur Fine, and Sheldon Goldstein. Dordrecht, The Netherlands, 1996. Pp. 211-229.

Berger, Peter L. and Thomas Luckmann. *The Social Construction of Reality.* Doubleday & Co., New York, NY, 2011 [1966].

Bernanke, Benjamin S. *The Courage To Act: A Memoir Of A Crisis And Its Aftermath.* Norton, New York, NY, 2015.

Black, Max. *Models And Metaphors*. Cornell University Press, Ithaca, NY, 1962.

Black, Donald. "The Purification of Sociology" in *Contemporary Sociology.* The American Sociological Association, Washington DC, 2000. Vol. 29, No. 5. (September, 2000) Pp. 704-709.

Blackmore, John T. *Ernst Mach: His Work, Life and Influence*. University of California Press, Los Angeles, CA, 1972.

Blaug, Mark. *Economic History and the History of Economics.* New York University Press, New York, NY, 1987.

Blinder, Alan S. *When the Music Stopped.* The Penguin Press, New York, NY, 2013.

Blinder, Alan S. and Mark Zandi. "How the Great recession was Brought to an End" in *Moody's Analytics* (July, 2010).

Bohm, David. "A Suggested Interpretation of the Quantum Theory in Terms of 'Hidden' Variables. I and II". *The Physical Review.* Vol. 85. (January 1952), Pp. 166-193.

Bohm, David. *Causality and Chance*. D. Van Nostrand, New York, NY, 1957.

Bohm, David. "A Proposed Explanation of Quantum Theory in Terms of Hidden Variables at a Sub-quantum Mechanical Level" in *Observation and Interpretation*. Edited by S. Korner. Butterworth Scientific Publications, London, 1957.

Bohm, David. "Classical and Nonclassical Concepts in the Quantum Theory: An Answer to Heisenberg's *Physics and Philosophy*". The British Journal for the Philosophy of Science. Vol. 12 (February 1962), Pp. 265-280.

Bohm, David. *Wholeness and the Implicate Order*. Routledge & Kegan Paul, London, England, 1980.

Bohm, David. "Hidden Variables and the Implicate Order" in *Quantum Implications: Essays in Honor of David Bohm*. Edited by B.J. Hiley and F. David Peat. Routledge & Kegan Paul, New York, NY, 1987. Pp. 33-45.

Bohm, David and F. David Peat. *Science, Order and Creativity.* Bantam Books, New York, NY, 1987.

Bohm, David and Basil J. Hiley. *The Undivided Universe: An Ontological Interpretation of Quantum Theory.* Routledge, New York, NY, 1993.

Bohr, Niels. *Atomic Theory and the Description of Nature.* Cambridge University Press, Cambridge, England, 1961 [1934].

Bohr, Niels. "Can Quantum Mechanical Description of Physical Reality Be Considered Complete?", *Physical Review*, Vol. 48 (1935), Pp. 696-702.

Bohr, Niels. "Discussion with Einstein" in *Albert Einstein: Philosopher-Scientist.* Edited by Paul A. Schilpp. Open Court, LaSalle, IL, 1949. Pp. 201-41.

Bohr, Niels. *Atomic Physics and Human Knowledge.* John Wiley & Sons, New York, NY, 1958.

Bohr, Niels. *Essays 1958-1962 on Atomic Physics and Human Knowledge.* Interscience Publishers, New York, NY, 1963.

Bohrnstedt, George W., David Knoke and Alisa Potter Mee. *Statistics for Social Data Analysis.* F.E. Peacock Publishers, Itasca, IL 2002.

Born, Max. "Physical Reality", *The Philosophical Quarterly.* Vol. 3 (1953), Pp. 139-149.

Born, Max. *My Life and My Views.* Charles Scribner's Sons, New York, NY, 1968.

Braithwaite, Richard B. *Scientific Explanation: A Study of the Function of Theory, Probability, and Law in Science.* Cambridge University Press, Cambridge, England, 1968 [1953].

Bridgman, Paul W. *The Logic of Modern Physics.* New York, NY 1927.

Brookes, Bertram C. "The Foundations of Information Science. Part I. Philosophical Aspects", *Journal of Information Science*, Vol. 2 (October 1980). Pp. 125-133.

Burns, Arthur F. *Wesley Clair Mitchell: The Economic Scientist.* National Bureau of Economic Research, New York, 1952.

Butterfield, Herbert. *The Origins of Modern Science 1300-1800.* MacMillan, New York, NY 1958.

Carnap, Rudolf. *The Logical Construction of the World.* Translated by R.A. George. The University of California Press, Berkeley, CA, 1967 [1928].

Carnap, Rudolf. "On the Character of Philosophical Problems" (1934) in *The Linguistic Turn.* Edited by Richard Rorty. The University of Chicago Press, Chicago, IL, 1967. Pp. 54-62.

Carnap, Rudolf. "Logical Foundations of the Unity of Science" in *International Encyclopedia of Unified Science*, Vol. I. Edited by Otto Carnap, Rudolf, Otto Neurath and Charles Morris. The University of Chicago Press, Chicago, IL, 1938. Pp. 42-62.

Carnap, Rudolf. "Foundations of Logic and Mathematics" in *International Encyclopedia of Unified Science*, Vol. I. Edited by Otto Neurath, Rudolf Carnap, and Charles Morris. The University of Chicago Press, Chicago, IL, 1938. Pp. 140-213.

Carnap, Rudolf. *Introduction to Semantics and Formalization of Logic*. Harvard University Press Cambridge, MA, 1958 [1942,1943].

Carnap, Rudolf. Meaning and Necessity: *A Study in Semantics and Modal Logic*. The University of Chicago Press, Chicago, IL, 1964 [1947].

Carnap, Rudolf. *Logical Foundations of Probability*. The University of Chicago Press, Chicago, IL, 1950.

Carnap, Rudolf. *The Continuum of Inductive Methods.* The University of Chicago Press, Chicago, IL, 1952.

Carnap, Rudolf and Yesohua Bar-Hillel. "Semantic Information", *British Journal for the Philosophy of Science*. Vol. 4 (1953). Pp. 147-157.

Carnap, Rudolf. "Meaning and Synonymy in Natural Languages" (1955) in *Meaning and Necessity: A Study in Semantics and Modal Logic*. The University of Chicago Press, Chicago, IL, 1964 [1947]. Pp. 233-247.

Carnap, Rudolf. "On Some Concepts of Pragmatics" (1956) in *Meaning and Necessity: A Study in Semantics and Modal Logic*. The University of Chicago Press, Chicago, IL, 1964 [1947]. Pp. 248-253.

Carnap, Rudolf. "The Methodological Character of Theoretical Concepts" in *The Foundations of Science And The Concepts of Psychology and Psychoanalysis*. Edited by Herbert Feigl and Michael Scriven. University of Minneapolis Press, Minneapolis, MN, 1956.

Carnap, Rudolf. "Intellectual Autobiography" in *The Philosophy of Rudolf Carnap*. Edited by Paul Arthur Schilpp. Open Court, LaSalle, IL, 1963, Pp. 3-84.

Carnap, Rudolf. *Philosophical Foundations of Physics: An Introduction to the Philosophy of Science.* Ed. Martin Gardner. Basic Books, New York, NY, 1966.

Carnap, Rudolf. *Studies in Inductive Logic and Probability*. Edited by Richard C. Jeffery. Vol. 1 (1971) and Vol. 2 (1980). University of California Press, Berkeley, CA.

Chester, Marvin. *Primer of Quantum Mechanics*. Dover Publications, New York, NY, 2003 [1992].

Chomsky, Noam. *Syntactic Structures.* Mouton de Gruyter, New York, NY, 2002 [1957].

Christ, Carl F. "History of the Cowles Commission 1932-1952" in *Economic Theory and Measurement: A Twenty-Year Research Report, 1932-1952.* Cowles Commission for Research in Economics, Chicago, IL, 1953.

Cohen, Morris R. and Ernest Nagel. *An Introduction to Logic And Scientific Method.* Harcourt, Brace, and World, New York, NY, 1934.

Commons, John R. *Institutional Economics: Its Place in Political Economy*. Two Volumes. University of Wisconsin Press, Madison, WI, 1959 [1934].

Commons, John R. *Economics of Collective Action*. University of Wisconsin Press, Madison, WI, 1970 [1950].

Conant, James B. *On Understanding Science: An Historical Approach*. Yale University Press, New Haven, CT, 1947.

Conant, James B. *Science and Common Sense*. Yale University Press, New Haven, CT, 1951.

Conant, James B. *Modern Science and Modern Man*. Columbia University Press, New York, NY, 1952.

Conant, James B. *My Several Lives: Memoirs of a Social Inventor*. Harper & Row, New York, NY, 1970.

Conant, James B. *Two Modes of Thought.* Trident Press, New York, NY, 1964.

Dahrendorf, Ralf. "Social Structure, Group Interests, and Conflict Groups" in *Class and Class Conflict in Industrial Society.* Stanford University Press, Palo Alto, CA, 1959.

Danielson, Dennis and Christopher M. Grancy. "The Case Against Copernicus" in *Scientific American.* Scientific American, Nature America, NY. Vol. 30, No. 1. (January 2014). Pp. 74-77.

Davis, Kingsley and Wilbert E. Moore. "Some Principles of Stratification", *American Sociological Review*. Vol. XXII (1957). Pp. 242-249.

De Saussure, Ferdinand. *Course in General Linguistics.* (Trans. Wade Baskin, Ed. Maun Saussy.) Columbus University Press, New York, NY, [1916] 2011.

Dillon, George L. *Introduction to Contemporary Linguistic Semantics.* Prentice-Hall, Englewood Cliffs, NJ, 1977.

Doan, Thomas, Robert Litterman, and Christopher Sims. "Forecasting and Conditional Projection Using Realistic Prior Distributions". *Research Department Staff Report 93.* Federal Reserve Bank of Minneapolis. July 1986. Also an earlier version in *Economic Review*. Vol. 3, No. 1. (1984). Pp. 1-100.

Dua, Pami, and Subhash C. Ray. "A BVAR Model for the Connecticut Economy", *Journal of Forecasting*. Vol. 14 (July 1994). Pp. 167-180.

Duhem, Pierre. *The Aim and Structure of Physical Theory.* Translated by Philip P. Wiener. Princeton University Press, Princeton, NJ, 1962 [1906].

Duhem, Pierre. *To Save the Phenomena: An Essay on the Idea of Physical Theory from Plato to Galileo.* Translated by Edmund Doland and Chaninah Maschler. The University of Chicago Press, Chicago, IL, 1969 [1908].

Ebenstein, Alan O. *Friedrich Hayek: A Biography.* Palgrave, New York, NY, 2001. See: "Postscript."

Eco, Umberto. *A Theory of Semiotics (Advances in Semiotics,) In*diana University Press, Bloomington. IN, 1976.
Eco, Umberto. *Semiotics and the Philosophy of Language (Advances in Semiotics).* Indiana University Press, Bloomington, IN, 1986.

Einstein, Albert. *Relativity: The Special and General Theory.* Crown Publishers, New York, NY, 1961[1916]. Translated by Robert W. Lawson.

Einstein, Albert. "On the Method of Theoretical Physics" in *Ideas and Opinions*. Crown Publishing, New York, 1954 [1933]. Pp. 270-274.

Einstein, Albert, B. Podolsky, and N. Rosen. "Can Quantum - Mechanical Description of Physical Reality Be Considered Complete?", *Physical Review*, Vol. 47 (May 1935), Pp. 777-780.

Einstein, Albert. "Physics and Reality", *The Journal of the Franklin Institute*, Vol. 221 (March 1936), Pp. 349-82. Translated by Jean Piccard.

Einstein, Albert. "Autobiographical Notes" in *Albert Einstein: Philosopher-Scientist.* Edited by Paul A. Schilpp. Open Court, LaSalle, IL, 1949. Pp. 2-96.

Einstein, Albert. "Remarks Concerning the Essays Brought Together in This Cooperative Volume" in *Albert Einstein:*

Philosopher-Scientist. Edited by Paul A. Schilpp. Open Court, LaSalle, IL, 1949. Pp. 665-688.

Feyerabend, Paul K. "On the Quantum-Theory of Measurement" in *Observation and Interpretation.* Edited by S. Korner. Butterworths Scientific Publications, London, England, 1957. Pp. 121-130.

Feyerabend, Paul K. "An Attempt at a Realistic Interpretation of Experience" in *Proceedings of the Aristotelian Society*. Vol. LVIII. Harrison & Sons, Ltd., London, England, 1958. Pp. 143-170.

Feyerabend, Paul K. "Complementarity" in *Aristotelian Society* (Supplementary Volume XXXII, 1958). Harrison and Sons, Ltd., London, England, 1958. Pp. 75-104.

Feyerabend, Paul K. "Problems of Microphysics" in *Frontiers of Science and Philosophy.* Edited by Robert G. Colodny. University of Pittsburgh Press, Pittsburgh, PA, 1962. Pp. 189-283.

Feyerabend, Paul K. "Explanation, Reduction and Empiricism" in *Minnesota Studies in the Philosophy of Science*. Edited by Herbert Feigl and Grover Maxwell. University of Minnesota Press, Minneapolis, MN, 1962. Pp. 28-97.

Feyerabend, Paul K. "Problems of Empiricism" in *Beyond the Edge of Certainty*. Edited by Robert G. Colodny. Prentice-Hall, Englewood Cliffs, NJ, 1965. Pp. 145-260.

Feyerabend, Paul K. "Problems of Microphysics" in *Philosophy of Science Today*. Edited by Sidney Morgenbesser. Basic Books, New York, NY, 1967. Pp. 136-147.

Feyerabend, Paul K. *Against Method: Outline of an Anarchistic Theory of Knowledge*. Verso, London, England, 1975.

Feyerabend, Paul K. *Science In A Free Society*. NLB, London, England, 1978.

Feyerabend, Paul K. *Realism, Rationalism, and Scientific Method: Philosophical Papers*. Vol. 1. Cambridge University Press, Cambridge, England, 1981.

Feyerabend, Paul K. *Problems of Empiricism: Philosophical Papers.* Vol. 2. Cambridge University Press, Cambridge, England, 1981.

Feyerabend, Paul K. *Farewell to Reason.* Verso, London, England, 1987.

Fine, Arthur. *The Shaky Game: Einstein, Realism, and the Quantum Theory.* Chicago University Press, Chicago, IL, 1996 (1986).

Form, William H. *Industrial Sociology*. Transaction Publishers, Harper Brothers, New York, NY, 1968 [1951].

Form, William H. "Institutional Analysis: An Organizational Approach" in *Change in Societal Institutions*. Edited by Maureen T. Hallinan *et al*. Plenum Press, New York, NY, 1996. Pp. 257-271.

Form, William H. *Work and Academic Politics: A Journeyman's Story*. Transaction Publishers, New Brunswick, NJ, 2002.

Fox, James A. *Forecasting Crime Data.* Lexington Books, Lexington, MA, 1978.

Freely, John. *Before Galileo: The Birth of Modern Science in Medieval Europe.* Overlook Duckworth, New York, NY, 2012.

Friedman, Milton. "The Methodology of Positive Economics" in *Essays in Positive Economics.* The University of Chicago Press, Chicago, IL, 1953. Pp. 3-43.

Friedman, Milton. *A Theory of the Consumption Function.* Princeton University Press, Princeton, NJ, 1957.

Gamow, George. *Thirty Years That Shook Physics: The Story of Quantum Theory.* Dover Publications, New York, NY, 1985 [1966].

Glaser, Barney G. and Anselm L. Strauss. *The Discovery of Grounded Theory: Strategies for Qualitative Research.* Aldine De Gruyter, New York, NY, 1967.

Glazer, Nathan. *The Limits of Social Policy.* Harvard University Press, Cambridge, MA, 1988.

Gouldner, Alvin W. *The Coming Crisis in Western Sociology.* Basic Books, New York, NY, 1970.

Grosser, Morton. *The Discovery of Neptune.* Harvard University Press, Cambridge, MA 1961.

Gruben, William C., and William T. Long III. "Forecasting the Texas Economy: Applications and Evaluation of a Systematic Multivariate Time-Series Model," *Economic Review,* Federal Reserve Bank of Dallas (January 1988).

Gruben, William C., and Donald W. Hayes. "Forecasting the Louisiana Economy," *Economic Review*, Federal Reserve Bank of Dallas (March 1991).

Gruben, William C., and William T. Long III. "The New Mexico Economy: Outlook for 1989," *Economic Review*, Federal Reserve Bank of Dallas (November 1988).

Gustavson, Carl G. *A Preface to History.* McGraw-Hill, New York, NY, 1955. P. 28.

Haavelmo, Trygve. "The Probability Approach in Econometrics", *Econometrica.* Vol. 12 (July 1944), Supplement.

Habermas, Jurgen. *On the Logic of the Social Sciences.* Translated by Shierry Weber Nicholsen and Jerry A. Stark, MIT Press, Cambridge, MA, 1988.

Hagstrom, Warren O. *The Scientific Community.* Basic Books, New York, NY, 1965.

Hanson, Norwood Russell. "On Elementary Particle Theory", *Philosophy of Science*, Vol. 23 (1956). Pp. 142-148.

Hanson, Norwood R. *Patterns of Discovery: An Inquiry into the Conceptual Foundations of Science.* Cambridge University Press, Cambridge, England, 1958.

Hanson, Norwood R. "Copenhagen Interpretation of Quantum Theory", *The American Journal of Physics*, Vol. 27 (1959). Pp. 1-15.

Hanson, Norwood R. "Niels Bohr's Interpretation of the Quantum Theory", *Current Issues in the Philosophy of Science.* Edited by Herbert Feigl and Grover Maxwell. Holt, Rinehart and Winston, New York, NY, 1961. Pp. 371-390.

Hanson, Norwood R. "Postscript" in *Quanta and Reality.* American Research Council, New York, NY, 1962. Pp. 85-93.

Hanson, Norwood R. *The Concept of the Positron.* Cambridge University Press, Cambridge, England, 1963.

Hanson, Norwood R. "Newton's First Law: A Philosopher's Door into Natural Philosophy" in *Beyond the Edge of*

Certainty. Edited by Robert G. Colodny. Prentice-Hall, Englewood Cliffs, NJ, 1965. Pp. 6-28.

Hanson, Norwood R. "Notes Toward A Logic Of Scientific Discovery" in *Perspectives on Peirce*. Edited by Richard J. Bernstein. Yale University Press, New Haven, CT, 1965. Pp. 43-65.

Hanson, Norwood R. "The Genetic Fallacy Revisited", *American Philosophical Quarterly*. Vol. 4 (April 1967), Pp. 101-113.

Hanson, Norwood R. "Quantum Mechanics, Philosophical Implications of" in *The Encyclopedia of Philosophy*. Vol. VII. Edited by Paul Edwards. The Macmillan Company & The Free Press, New York, NY, 1967. Pp. 41-49.

Hanson, Norwood R. "Observation and Interpretation" in *Philosophy of Science Today*. Edited by Sidney Morgenbesser. Basic Books, New York, NY, 1967. Pp. 89-99.

Hanson, Norwood R. *Perception and Discovery: An Introduction to Scientific Inquiry*. Edited by Willard C. Humphreys. Freeman, Cooper & Company, San Francisco, CA, 1969.

Hanson, Norwood R. *Observations and Explanations: A Guide to Philosophy of Science.* Harper & Row, Publishers, New York, NY, 1973.

Hayek, Frederich A. *The Counter-Revolution of Science: Studies on the Abuse of Reason.* Free Press, Glencoe, IL, 1955.

Heelan, Patrick A., S.J. *Quantum Mechanics and Objectivity: A Study of the Physical Philosophy of Werner Heisenberg.* Martinus Nijhoff, The Hague, Netherlands, 1965.

Heisenberg, Werner. *The Physical Principles of the Quantum Theory.* Translated by Carl Eckart and F. C. Hoyt. Dover Publications, Inc., New York, NY, 1950 [1930].

Heisenberg, Werner. *The Physicist's Conception of Nature.* Translated by Arnold J. Pomerans. Harcourt, Brace and Company, New York, NY 1955.

Heisenberg, Werner. "The Development of the Interpretation of the Quantum Theory" in *Niels Bohr and The Development of The Quantum Physics*. Edited by Wolfgang Pauli. McGraw-Hill, New York, NY, 1955.

Heisenberg, Werner. *Physics and Philosophy: The Revolution in Modern Science*. Harper and Row, New York, NY, 1958.

Heisenberg, Werner, Max Born, Erwin Schrödinger, and Pierre Auger. *On Modern Physics.* Collier Books, New York, NY, 1961.

Heisenberg, Werner. "Quantum Theory and Its Interpretation" in *Niels Bohr*. Edited by S. Rosental. Interscience Press, New York, NY, 1964. Pp. 94-108.

Heisenberg, Werner. *Physics and Beyond: Encounters and Conversations.* Translated by Arnold J. Pomerans. Harper and Row, New York, NY, 1971.

Heisenberg, Werner. *Across the Frontiers*. Translated by Peter Heath. Harper and Row, New York, NY, 1974.

Heisenberg, Werner. *Philosophical Problems of Quantum Physics*. Translated by F. C. Hayes. Ox Bow Press, Woodbridge, CT, 1979. Formerly titled *Philosophic Problems of Nuclear Science*. Pantheon, New York, NY, 1952.

Heisenberg, Werner. “Remarks on the Origin of the Relations of Uncertainty” in *The Uncertainty Principle and Foundations of Quantum Mechanics*. Edited by William C. Price and Seymour S. Chissick. John Wiley & Sons, New York, NY, 1977.

Hempel, Carl G. and Paul Oppenheim. “Logic of Explanation”, *Philosophy of Science*. Vol. 16 (1948). Pp. 135-175.

Hempel, Carl G. *Philosophy of Natural Science*. Prentice-Hall, Englewood Cliffs, NJ, 1966.

Hendry, David F. “Modelling UK Inflation, 1875-1991”, *Journal of Applied Econometrics*. Vol. 16 (2001). Pp. 255-273.

Hendry, David F. and Jurgen A. Doornik. *Empirical Model Discovery and Theory Evaluation: Automatic Selection Methods in Econometrics*. The MIT Press, Cambridge, MA 2014.

Henry, D.P. *Medieval Logic & Metaphysics*. Hutchinson, London, England, 1972.

Hesse, Mary B. *Models and Analogies in Science*. Sheed and Ward, London, England, 1953.

Hesse, Mary B. “Models and Matter” in *Quanta and Reality*. American Research Council, New York, NY, 1962. Pp. 49-60.

Hesse, Mary B. *Forces and Fields: The Concept of Action at a Distance in the History of Physics.* Greenwood Press, Westport, CT 1962.

Hesse, Mary B. “The Explanatory Function of Metaphor” in *Logic, Methodology, and Philosophy of Science*. Edited by Y. Bar-Hillel. Amsterdam, Holland, 1965.

Hesse, Mary B. "Models and Analogy in Science" in *The Encyclopedia of Philosophy*. Vol. VII. Edited by Paul Edwards. The Macmillan Company & The Free Press, New York, NY, 1967. Pp. 354-359.

Hesse, Mary B. "Laws and Theories" in *The Encyclopedia of Philosophy*. Vol. VII. Edited by Paul Edwards. The Macmillan Company & The Free Press, New York, NY, 1967. Pp. 404-410.

Hickey, Thomas J. *Introduction to Metascience: An Information Science Approach to Methodology of Scientific Research*. Hickey, Oak Park, IL 1976.

Hickey, Thomas J. "The Indiana Economic Growth Model", *Perspectives on the Indiana Economy.* Vol. 2, (March 1985). Pp. 1-11.

Hickey, Thomas J. "The Pragmatic Turn in the Economics Profession and in the Division of Economic Analysis of the Indiana Department of Commerce", *Perspectives on the Indiana Economy.* Vol. 2, (September 1985). Pp. 9-16.

Hickey, Thomas J. "Understanding Feyerabend on Galileo," *Irish Theological Quarterly,* Vol. 74, No. 1. (February 2009), 89-92.

Hickey, Thomas J. *Twentieth-Century Philosophy of Science: A History.* Hickey, River Forest, IL. e-book [2016, (1995)].

Hickey, Thomas J. *Philosophy of Science: An Introduction* (Fifth Edition), e-book, 2019.

Hoehn, James G., and James J. Balazsy. "The Ohio Economy: A Time-Series Analysis," *Economic Review*, Federal Reserve Bank of Cleveland (Third Quarter, 1985).

Hoehn, James G., William C. Gruben, and Thomas B. Fomby. "Time Series Forecasting Models of the Texas Economy: A Comparison," *Economic Review*, Federal Reserve Bank of Dallas (May 1984).

Holyoak, Keith and Paul Thagard. *Mental Leaps: Analogy in Creative Thought*. The MIT Press/Bradford Books, Cambridge, England, 1995.

Humphreys, Paul. *Extending Ourselves.* Oxford University Press, New York, NY, 2004.

Jammer, Max. *The Conceptual Development of Quantum Mechanics*. McGraw-Hill, New York, NY, 1966.

Jammer, Max. *The Philosophy of Quantum Mechanics*. John Wiley & Sons, New York, NY 1974.

Keith, Bruce and Nicholas Babchuk. "The Quest for Institutional recognition: A Longitudinal Analysis of Scholarly Productivity and Academic Prestige among Sociology Departments", *Social Forces*, University of North Caroline Press, (June 1998).

Keynes, John Maynard. *The General Theory of Employment, Interest, and Money*. Harcourt, Brace & World, New York, NY 1964 [1936].

Kluback, William. *Wilhelm Dilthey's Philosophy of History.* Columbia University Press, New York, NY, 1956.

Klein, Lawrence R. *The Keynesian Revolution*. MacMillan, New York, 1966, [1947].

Kneale, William and Martha Kneale. *The Development of Logic*. Clarendon Press, Oxford, England, 1962.

Kuhn, Thomas S. *The Copernican Revolution: Planetary Astronomy in the Development of Western Thought.* Harvard University Press, Cambridge, MA, 1957.

Kuhn, Thomas S. *The Structure of Scientific Revolutions*. The University of Chicago Press, Chicago, IL, 1970 [1962].

Kuhn, Thomas S. "The Function of Dogma in Scientific Research" in *Scientific Change*. Edited by A.C. Crombie. Basic Books, New York, NY, 1963.

Kuhn, Thomas S. "Logic of Discovery or Psychology of Research?" in *Criticism and the Growth of Knowledge*. Edited by Imre Lakatos and Alan Musgrave. Cambridge University Press, Cambridge, England, 1970. Pp. 1-23.

Kuhn, Thomas S. "Reflections on My Critics" in *Criticism and the Growth of Knowledge*. Edited by Imre Lakatos and Alan Musgrave. Cambridge University Press, Cambridge, England, 1970. Pp. 231-278.

Kuhn, Thomas S. *The Essential Tension.* University of Chicago Press, Chicago, IL, 1977.

Kuhn, Thomas S. *Black-Body Theory and the Quantum Discontinuity* 1894-1912. The Clarendon Press, Oxford, England, 1978.

Kuhn, Thomas S. *The Road Since Structure: Philosophical essays, 1970-1993.* Edited by James Conant and John Haugeland. Chicago University Press, Chicago, IL, 2000.

Kulp, Christopher B. *Realism/Antirealism and Epistemology*. Rowman and Littlefield, New York, NY, 1997.

Kuprianov, Anatoli, and William Lupoletti. "The Economic Outlook for Fifth District States in 1984: Forecasts from Vector

Autoregression Models", *Economic Review*, Federal Reserve Bank of Richmond (February 1984).

Land, Kenneth C. "A Mathematical Formalization of Durkheim's Theory of their Causes of the Division of Labor" in *Sociological Methodology*. Edited by E.F. Borgatta and G.W. Borhnstedt. Jossey-Bass, San Francisco, CA, 1970.

Land, Kenneth C. "Social Indicators" in *Social Science Methods*. Edited by R.B. Smith. The Free Press, New York, 1971.

Land, Kenneth C. "Formal Theory" in *Sociological Methodology*. Edited by H.L. Costner. Jossey-Bass, San Francisco, CA, 1971.

Land, Kenneth C. "Theories, Models and Indicators of Social Change", *International Social Science*. Vol. 27, No. 1 (1975), Pp. 7-37.

Land, Kenneth C. and Marcus Felson. "A General Framework for Building Dynamic Macro Social Indicator Models", *American Journal of Sociology*. Vol. 82, No. 3 (Nov. 1976), Pp. 565-604.

Land, Kenneth C. "Modeling Macro Social Change'", *Sociological Methodology*. Vol. 11, (1980), Pp. 219-278.

Land, Kenneth C. "Social Indicators", Annual *Review of Sociology*. Vol. 9, (1983), Pp. 1-26.

Land, Kenneth C. "Models and Indicators", *Social Forces*. Vol. 80, No. 3 (Dec. 2001), Pp. 381-410.

Land, Kenneth C., Alex Michalos and M. Joseph Sirgy. *Handbook of Social Indicators and Quality of Life Research.* Springer, New York, NY, 2012.

Landé, Alfred. "Probability in Classical and Quantum Physics" in *Scientific Papers Presented to Max Born.* Hafner Publishing Company, New York, NY, 1954.

Landé, Alfred. *Foundations of Quantum Theory: A Study in Continuity and Symmetry.* Yale University Press, New Haven, CT, 1955.

Landé, Alfred. *New Foundations of Quantum Physics.* Cambridge University Press, Cambridge, England, 1965.

Langley, Pat and Jeff Shrager. *Computational Models of Scientific Discovery and Theory Formation.* Morgan Kaufmann, San Mateo, CA, 1990.

Langley, Pat, Saito Kazumi, George Deleep, Stephen Bay and Kevin R. Arrigo. "Discovering Ecosystem Models from Time-Series Data" in *Proceedings from the Six International Conference on Discovery Science.* Springer, Saporro, Japan, 2003. Pp. 141-152.

Langley, Pat and Will. Bridewell. "Processes and Constraints in Explanatory Scientific Discovery" in *Proceedings of the Thirteenth Annual Meeting of the Cognitive Science Society.* Washington, D.C. 2008.

Langley, Pat and Will. Bridewell. "Two Kinds of Knowledge in Scientific Discovery" in *Topics in Cognitive Science.* Vol. 2 (2010), Pp. 36-52.

Langley, Pat, Chunki Park and Will. Bridewell. "Integrated Systems for Inducing Spatio-Temporal Process Models" in

Fourth AAAI Conference on Artificial Intelligence. AAAI Press, Atlanta, GA, 2010. Pp. 1555-1560.

Langley, Pat, L. Todorovki and Will. Bridewell. "Discovering Constraints for Inductive Process Modeling" in *Proceedings of the Twenty-Sixth AAAI Conference on Artificial Intelligence*. AAAI Press, Toronto, Canada, 2012. Pp. 256-262.

Lekachman, Robert (ed). *Keyned and the Classics.* D.C. Heath & Co., Boston, MA, 1964.

Leplin, Jarrett. *A Novel Defense of Scientific Realism.* Oxford University Press, New York, NY, 1997.

Levenson, Thomas. *The Hunt for Vulcan and How Albert Einstein Destroyed a Planet, Discovered Relativity and Deciphered the Universe.* Random House, New York, NY, 2015.

Lipson, Hod and Michael Schmidt. "Distilling Free-Form Natural Laws from Experimental Data", *Science*. Vol. 324 (April 2009), Pp. 81-85.

Litterman, Robert B. *Techniques for Forecasting Using Vector Autoregressions.* Federal Reserve Bank of Minneapolis, Minneapolis, MN, 1979.

Litterman, Robert B. *Specifying Vector Autoregressions for Macroeconomic Forecasting.* Federal Reserve Bank of Minneapolis, Minneapolis, MN, 1984.

Litterman, Robert B. "Above-Average National Growth in 1985 and 1986", *Federal Reserve Bank of Minneapolis Quarterly Review.* Vol. 8 (Fall, 1984), Pp. 3-7.

Litterman, Robert B. *Forecasting with Bayesian Vector Autoregressions - Four Years of Experience.* Federal Reserve Bank of Minneapolis, Minneapolis, MN 1985.

Litterman, Robert B. "How Monetary Policy in 1985 Affects the Outlook", *Federal Reserve Bank of Minneapolis Quarterly Review*. Vol. 9 (Fall, 1985), Pp. 2-13.

Lucas, Robert E. "Econometric Policy Evaluation: A Critique" in *Phillips Curve and Labor Markets*. Edited by K. Brunner and A. H. Meltzer. North Holland, Amsterdam, The Netherlands, 1976.

Lundberg, George A. "Contemporary Positivism in Sociology", American *Sociological Review.* Vol. 4 (February 1939). Pp. 42-55.

Lundberg, George A. *Foundations of Sociology*. Greenwood Press, Westport, CT, 1964 [1939].

Lundberg, George A. *Social Research.* Longmans, Green & Co., New York, NY, 1942.

Lundberg, George A. C*an Science Save Us?* David McKay, New York, NY, 1961 [1947].

Mach, Ernst. *The Analysis of Sensations and the Relation of the Physical to the Psychical.* Translated by C. M. Williams and Sydney Waterlow. Dover Publications, New York, NY, 1959 [1885].

Mach, Ernst. *The Science of Mechanics: A Critical and Historical Account of Its Development*. Translated by Thomas J. McCormack. Open Court, LaSalle, IL, 1960 [1893].

Mach, Ernst. *Popular Science Lectures*. Translated by Thomas J. McCormack. Open Court, LaSalle, IL, 1943 [1898].

Mach, Ernst. *The Principles of Physical Optics: An Historical and Philosophical Treatment.* Translated by John S. Anderson and A.F.A. Young. Dover, New York, NY, 1953 [1921].

MacIver, Robert M. *Social Causation.* Peter Smith, Gloucester, MA, 1973.

Mates, Benson. *Stoic Logic.* University of California Press, CA, 1973.

Maus, Heinz. *A Short History of Sociology.* Philosophical Library, New York, NY 1962.

McNeill, William H. *Plagues and Peoples.* Doubleday, New York, NY, 1977.

Merton, Robert K. "The Unanticipated Consequences of Purposive Action." *American Sociological Review*. Vol.1, No. 6 (December 1936). Pp. 894-904.

Merton, Robert K. *Social Theory and Social Structure.* The Free Press, New York, NY, 1968 [1947].

Merton, Robert K. *Sociology of Science: Theoretical and Empirical Investigations*. University of Chicago Press, Chicago, IL, 1973.

Miceli, G. and Caramazza, Alfonso. "The Assignment of Word Stress: Evidence from a Case of Acquired Dyslexia." *Cognitive Neuropsychology*. Vol. 10, No. 3 (1993), Pp. 113-141.

Michalos, Alex C. "Positivism versus the Hermeneutic-Dialectic School", *Theoria.* Vol. 35 (1969) Pp. 267-278.

Michalos, Alex C. "Philosophy of Social Science" in *Current Research in Philosophy of Science*. Edited by Peter D. Asquith and Henry E. Kyburg. Philosophy of Science Association, East Lansing, MI 1979.

Michalos, Alex C. “Philosophy of Science: Historical, Social and Value Aspects” in *A Guide to the Culture of Science, Technology, and Medicine.* Edited by Paul T. Durbin. The Free Press, New York, NY, 1980.

Mitchell, Wesley C. *Business Cycles*. University of California Press. Los Angeles, CA, 1913. Part III published as *Business Cycles and Their Causes*. 1941.

Mitchell, Wesley C. “The Prospects of Economics” in *The Trend of Economics*. Vol. I. Edited by Rexford G. Tugwell. Kennikat Press, Port Washington, NY, 1971 [1924]. Pp. 3-31.

Mitchell, Wesley C. “Quantitative Analysis in Economic Theory”, *American Economic Review.* Vol. 15 (1925), Pp. 1-12.

Mitchell, Wesley C. *Measuring Business Cycles*. National Bureau of Economic Research, New York, 1946.

Mitchell, Wesley C. *Types of Economic Theory*. Vols. I and II. Edited by Joseph Dorfman. Augustus M. Kelley, New York, NY 1967.

Morgan, Mary S. *The History of Econometric Ideas*. Cambridge University Press, Cambridge, England. 1990.

Muth, John F. “Optimal Properties of Exponentially Weighted Forecasts”, *Journal of the American Statistical Association,* Vol. 55 (1960), Pp. 299-306.

Muth, John F. “Rational Expectations and the Theory of Price Movements” *Econometrica*. Vol. 29 (1961) Pp. 315-335.

Myrdal, Gunnar. *Against the Stream: Critical Essays on Economics.* Pantheon Books, New York, NY, 1973.

Nagel, Ernest. *The Structure of Science*. Harcourt, Brace, World, New York, NY, 1961.

National Science Foundation. *Science and Engineering Doctorates: 1960-1988*. Washington, DC, 1989.

Neurath, Otto. "Foundations of the Social Sciences" in I*nternational Encyclopedia of Unified Science,* Vol. II. Edited by Otto Neurath, Rudolf Carnap and Charles Morris. University of Chicago Press, Chicago, IL, 1944.

Neurath, Otto. *Empiricism and Sociology*. Edited by Marie Neurath and Robert S. Cohen. D. Reidel, Dordrecht, Holland, 1973.

Ohlin, Bertil G. "The Stockholm Theory of Savings and Investment", *Economic Journal*, Vol. 47 (March 1937), Pp. 221-240.

Ohlin, Bertil G. *The Problem of Employment Stabilization.* Greenwood Press, Westport, CT, 1977 [1949].

Omnès, Roland. *The Interpretation of Quantum Mechanics.* Princeton University Press, Princeton, NJ, 1994.

Omnès, Roland. *Understanding Quantum Mechanics.* Princeton University Press, Princeton, NJ, 1999.

Orenstein, Alex. *Willard Van Orman Quine*. Twayne, Boston, MA, 1977.

Palmer, Richard E. *Hermeneutics: Interpretation Theory in Scheiermacher, Dilthey, Heidegger and Gadamer.* Northwestern University Press, Evanston, IL, 1969.

Parsons, Talcott. *The Structure of Social Action.* The Free Press, New York, IL, 1968, [1937]. Vols. I and II.

Parsons, Talcott. *The Social System.* The Free Press, New York, IL, 1964, [1951].

Parsons, Talcott. "On Building Social Systems Theory" in *The Twentieth-Century Sciences: Studies in the Biography of Ideas.* Edited by Gerald Holton. W.W. Norton, New York, NY, 1972 [1970]. Pp. 99-154.

Peat, F. David. *Infinite Potential: The Life and Times of David Bohm.* Addison-Wesley Publishing Co., New York, NY, 1997.

Peirce, Charles S. *Philosophical Writings of Peirce*. Edited and Selected by Justice Buchler. Dover, NY, 1955 [1940].

Peirce, Charles S. *Collected Papers*. Edited by C. Hartsorne and P. Weiss. Cambridge, MA, 1934.

Peirce, Charles S. *Essays in the Philosophy of Science*. Edited with introduction by Vincent Thomas. Bobbs-Merrill, New York, NY, 1957.

Planck, Max. *Scientific Autobiography and Other Papers*. Philosophical Library, New York, NY, 1949.

Poole, Stuart C. *An Introduction to Linguistics.* Saint Martin's Press, New York, NY, 1999.

Popper, Karl R. *The Open Society and its Enemies*. Princeton University Press, Princeton, NJ 1950.

Popper, Karl R. *The Poverty of Historicism*. Harper and Row, New York, NY, 1961 [1957].

Popper, Karl R. *Conjectures and Refutations*: The Growth of Scientific Knowledge. Basic Books, New York, NY, 1963.

Popper, Karl R. *The Logic of Scientific Discovery*. Harper and Row, New York, NY, 1968 [1959,1934].

Popper, Karl R. "Normal Science and its Dangers" in *Criticism and the Growth of Knowledge*. Edited by Imre Lakatos and Alan Musgrave. Cambridge University Press, Cambridge, England, 1970. Pp. 51-58.

Popper, Karl R. "Autobiography of Karl Popper" in *The Philosophy of Karl Popper*. Edited by Paul A. Schilpp. Two Vols. Open Court, LaSalle, IL, 1972. Pp. 3-181.

Popper, Karl R. *Objective Knowledge: An Evolutionary Approach*. Clarendon Press, Oxford, England, 1974.

Popper, Karl R. "The Rationality of Scientific Revolutions" in *Problems of Scientific Revolution: Progress and Obstacles to Progress in the Sciences*. Edited by Rom Harre. Clarendon Press, Oxford, England, 1975.

Popper, Karl R. and John C. Eccles. *The Self and Its Brain*. Springer International, New York, NY, 1977.

Popper, Karl R. *Realism and the Aim of Science.* Edited by W.W. Bartley. Rowman and Littlefield, Totowa, NJ, 1983.

Popper, Karl R. *The Open Universe.* Edited by W.W. Bartley, III. Rowman and Littlefield, Totowa, NJ, 1983.

Popper, Karl R. *Quantum Theory And The Schism in Physics*. Edited by W.W. Bartley. Rowman and Littlefield, Totowa, NJ, 1983.

Putnam, Hilary. Reason, Truth and History. Cambridge University Press, New York, NY. 1981.

Putnam, Hilary. *Representation and Reality*. MIT Press, Cambridge, MA 1988.

Quine, W.V.O. *Methods of Logic.* Harvard University Press, Cambridge, MA, 1982 [1950].

Quine, W.V.O. *From A Logical Point of View*. Harvard University Press, Cambridge, MA, 1980 [1953].

Quine, W.V.O. *Word and Object*. M.I.T. Press, Cambridge, MA, 1960.

Quine, W.V.O. *Selected Logic Papers*. Random House, New York, NY, 1966.

Quine, W.V.O. "Linguistics and Philosophy" in *Language and Philosophy*. Edited by S. Hook. New York University Press, New York, NY, 1969.

Quine, W.V.O. *Ontological Relativity*. Columbia University Press, New York, NY, 1969.

Quine, W.V.O. *Philosophy of Logic*. Harvard University Press, Cambridge, MA, 1970.

Quine, W.V.O., and J.S. Ullian. *The Web of Belief*. Random House, New York, NY, 1970.

Quine, W.V.O. *The Roots of Reference*. Open Court, LaSalle, IL, 1974.

Quine, W.V.O. "Methodological Reflections on Current Linguistic Theory" in On *Noam Chomsky*. Edited by G. Harmon. Doubleday, Garden City, NY, 1974.

Quine, W.V.O. *Theories and Things*. Belknap Press, Cambridge, MA, 1981.

Quine, W.V.O. "Autobiography of W. V. Quine" in *The Philosophy of W.V. Quine*. Edited by Lewis Edwin Hahn and Paul Arthur Schilpp. Open Court, LaSalle, IL, 1986, Pp. 3-46.

Quine, W.V.O. *The Time of My Life*: An Autobiography. MIT Press, Cambridge, MA, 1985.

Quine, W.V.O. *Quiddities: An Intermittently Philosophical Dictionary*. Belknap Press, Cambridge, MA, 1987.

Quine, W.V.O. and Rudolf Carnap. *Dear Carnap, Dear Van: The Quine-Carnap Correspondence and Related Works*. Edited with introduction by Richard Creath. University Of California Press, Berkeley, CA, 1990.

Radnitzky, Gerald. *Contemporary Schools of Metascience.* Vols. I and II. Akademiforlagst, Goteborg, Sweden, 1968.

Redman, Deborah A. *Economic Methodology: A Bibliography with References to Works in the Philosophy of Science*. Complied by Deborah A. Redman. Greenwood Press, Westport, CT, 1989.

Rogers, Rolf E. *Max Weber's Ideal Type Theory*. Philosophical Library, New York. NY, 1969.

Rorty, Richard M. (ed.) *The Linguistic Turn: Essays in Philosophical Method.* University of Chicago Press, Chicago, IL, [1967], 1992.

Rosenberg, Alexander. *Philosophy of Social Science*. Westview Press, Boulder, CO, 1995.

Roseveare, N.T. *Mercury's Perihelion: From LeVerrier to Einstein.* Clarendon Press, Oxford, England, 1982.

Samuelson, Paul A. "Interaction between the Multiplier and the Principle of Acceleration", *Review of Economics and Statistics*. 21(2) 1938. Pp. 75-78.

Samuelson, Paul A. "Lord Keynes and the General Theory" *Econometrica*. Vol XIV, (July 1946) P. 187.

Samuelson, Paul A. 1966. "A Note on the Pure Theory of Consumer's Behavior" (1938) and "Consumption Theory in

Terms of Revealed Preference" (1948). *The Collected Papers of Paul A. Samuelson,* Vol. 1.

Samuelson, Paul A. and Robert Lekachman (ed.) *Keynes General Theory: Reports of Three Decades*. St. Martin's Press, NY, 1964.

Samuelson, Paul A. *Economics.* McGraw-Hill, NY, 1973.

Sapir, Edward. *Language.* Rupert-Hart Davis, London, 1963.

Sardar, Ziauddin. Thomas Kuhn and the Science Wars. Totem Books, New York, NY, 2000.

Sargent, Thomas J. "Rational Expectations, Econometric Exogeniety, and Consumption", Journal *of Political Economy*. Vol. 86 (August 1978), Pp. 673-700.

Sargent, Thomas J. "Estimating Vector Autoregressions Using Methods Not Based on Explicit Economic Theories", *Federal Reserve Bank of Minneapolis Quarterly Review,* Vol. 3. (Summer 1979). Pp. 8-15.

Sargent, Thomas J. "After Keynesian Macroeconomics" in *Rational Expectations and Econometric Practice.* Edited by Robert E. Lucas and Thomas J. Sargent. University of Minnesota Press, Minneapolis, MN, 1981.

Schrödinger, Erwin. *Science and the Human Temperament.* Translated by James Murphy and W.H. Johnston. W.W. Norton & Company, New York, NY, 1935 [German, 1932].

Schrödinger, Erwin. *My View of the World.* Translated by Cecily Hastings. Ox Bow Press, Woodbridge, CT, 1983 (1961).

Schumpeter, Joseph A. *Essays on Economic Topics of Joseph A. Schumpeter.* Edited by Richard V. College. Kennikat Press, Port Washington, NY, 1951.

Schumpeter, Joseph A. *History of Economic Analysis.* Oxford Univer4sity Press, Fair Lawn, NJ, 1954.

Searle, John R. "Chomsky's Revolution in Linguistics" in On *Noam Chomsky.* Edited by G. Harmon. Doubleday, Garden City, NY, 1974.

Searle, John R. "Does the Real World Exist?" in *The Construction of Social Reality.* The Free Press, New York, NY, 1995.

Searle, John R. *The Construction of Social Reality.* The Free Press, New York, NY, 1995.

Searle, John R. *Mind, Language and Society: Philosophy in the Real World.* Basic Books, New York, NY, 1998.

Sedelow, Walter A. and Sally Yeates Sedelow. "The History of Science as Discourse, The Structure of Scientific and Literary Texts", *Journal of the History of the Behavioral Sciences*. Vol. 15, No. 1 (January 1979). Pp. 63-72.

Shannon, Claude E. and Warren Weaver. *The Mathematical Theory of Communication*. The University of Illinois Press, Urbana, IL, 1949.

Shapere, Dudley. "The Structure of Scientific Revolutions", *Philosophical Review*, Vol. 73 (1964), Pp. 383-94.

Shiller, Robert J. *Irrational Exuberance.* Princeton University Press, Princeton, NJ, 2015.

Shreider, Yu A. "On the Semantic Characteristics of Information" in *Information Storage and Retrieval.* Vol. II (August 1965).

Shreider, Yu A. "Semantic Aspects of Information Theory" in *On Theoretical Problems On Informatics.* All-Union Institute for Scientific and Technical Information, Moscow, USSR, 1969.

Shreider, Yu A. "Basic Trends in the Field of Semantics" in *Statistical Methods in Linguistics.* Sprjakfhorlaset Skriptor, Stockholm, Sweden, 1971.

Simon, Herbert A. T*he Sciences of the Artificial.* MIT Press, Cambridge, MA, 1969.

Simon, Herbert A. and Laurent Silkossy. *Representation and Meaning with Information Processing Systems.* Prentice-Hall, Englewood Cliffs, NJ, 1972.

Simon, Herbert A. *Models of Discovery and Other Topics in the Methods of Science*. D. Reidel, Dordrecht, Holland, 1977.

Simon, Herbert A., et al. *Scientific Discovery: Computational Explorations of the Creative Process*. The MIT Press, Cambridge, MA, 1987.

Simon, Herbert A. *Models of My Life*. Basic Books, Dunmore, PA, 1991.

Sims, Christopher A. "Macroeconomics and Reality" *Econometrica*. Vol. 48. (January 1980), Pp. 1-47.

Sims, Christopher A. "Are Forecasting Models Usable for Policy Analysis?", Federal *Reserve Bank of Minneapolis Quarterly Review.* Vol. 10 (Winter 1986), Pp. 2-16.

Sonquist, John A. "Problems in the Analysis of Survey Data, and A Proposal", *Journal of the American Statistical Association*. Vol. 58, (June 1963), Pp. 415-35.

Sonquist, John A. and James N. Morgan. *The Detection of Interaction Effects: A Report on a Computer Program for the Selection of Optimal Combinations of Explanatory Variables.* Survey Research Center, Institute for Social Research, University of Michigan, Ann Arbor, MI, 1964.

Sonquist, John A. "Simulating the Research Analyst" in *Social Science Information*. Vol. VI, No 4 (1967), Pp. 207-215.

Sonquist, John A. *Multivariate Model Building: Validation of a Search Strategy*. Survey Research Center, Institute for Social Research, University of Michigan, Ann Arbor, MI, 1970.

Sonquist, John A., Elizabeth L. Baker, and James N. Morgan. *Searching for Structure: An Approach to Analysis of Substantial Bodies of Micro-Data and Documentation for a Computer Program.* Survey Research Center, Institute for Social Research, University of Michigan, Ann Arbor, MI, 1973.

Sonquist, John A. "Computers and the Social Sciences", *American Behavioral Scientist*. Vol. 20, No. 3 (1977), Pp. 295-318.

Sonquist, John A. and Francis M. Sim. "'Retailing' Computers to Social Scientists" *American Behavioral Scientist*. Vol. 20, No. 4 (1977), Pp. 319-45.

Soros, George. *The Alchemy of Finance: Reading the Mind of the Market.* John Wiley & Sons, New York, NY, 1994 [1987].
Stiglitz, Joseph E. *Freefall.* W.W. Norton & Company, New York, NY, 2010.

Stone, Richard. *Demographic Accounting and Model-Building*. Organization for Economic Co-operation and Development, Paris, 1971.

Stiglitz, Joseph E. *Freefall.* W.W. Norton & Company, New York, NY, 2010.

Stein, Maurice and Arthur Vidich. *Sociology on Trial.* Prentice-Hall, Englewood Cliffs, NJ, 1963.

Tainter, Joseph A. *The Collapse of Complex Societies.* Cambridge University Press, New York, NY 1988.

Tarski, Alfred. *Logic, Semantics, Metamathematics.* Trans. by J.H. Woodger. Clarendon Press, Oxford, England, 1956.
Thagard, Paul. "The best explanation: Criteria for theory choice." *Journal of Philosophy*. Vol. 75 (1978), Pp. 76-92.

Thagard, Paul and Keith Holyak. "Discovering The Wave Theory of Sound: Inductive Inference in the Context of Problem Solving" in *Proceedings of the Ninth International Joint Conference on Artificial Intelligence*. Morgan Kaufmann, Los Angeles, CA, 1985. Vol. I, Pp. 610-12.

Thagard, Paul. *Computational Philosophy of Science*. MIT Press/Bradford Books, Cambridge, MA, 1988.

Thagard, Paul. "Explanatory Coherence." *Behavioral and Brain Sciences*. Vol. 12 (1989), Pp. 435-467.

Thagard, Paul, D. Cohen and K. Holyoak. "Chemical Analogies: Two Kinds of Explanation" in *Proceedings of the Eleventh International Joint Conference on Artificial Intelligence*. Morgan Kaufmann, San Mateo, CA, 1989.

Thagard, Paul. The dinosaur debate: Explanatory coherence and the problem of competing hypotheses" in J. Pollock and R. Cummins (Eds.), *Philosophy and AI: Essays at the Interface*. MIT Press/Bradford Books, Cambridge, MA, 1991.

Thagard, Paul and Greg Nowak. "Copernicus, Ptolemy, and Explanatory Coherence" in *Minnesota Studies in the Philosophy of Science: Cognitive Models of Science*. (Ed. Ronald N. Giere) University of Minnesota Press, Minneapolis, MN 1992. Vol. XV, Pp. 274-309.

Thagard, Paul. *Conceptual Revolutions*. Princeton University Press, Princeton, NJ, 1992.

Thagard, Paul. *Mind: Introduction to Cognitive Science*. The MIT Press/Bradford Books, Cambridge, MA, 1996.

Thagard, Paul. *How Scientists Explain Disease*. Princeton University Press, Princeton, NJ, 1999.

Thagard, Paul *et al*. *Model-Based Reasoning in Scientific Discovery*. (Edited by Lorenzo Magnani, Nancy J. Neressian, and Paul Thagard). Kluwer Academic/Plenum Publishers, New York, NY, 1999.

Thaler, Richard H. *Misbehaving: The Making of Behavioral Economics.* W.W. Norton, New York, NY, 2015.

Todd, Richard M. "Improving Economic Forecasting with Bayesian Vector Auto-regression," *Quarterly Review*, Federal Reserve Bank of Minneapolis, (Fall 1984).

Todd, Richard M. and William Roberds. "Forecasting and Modeling the U.S. Economy", F*ederal Reserve Bank of Minneapolis Quarterly Review.* Vol. 11. (Winter 1987), Pp. 7-20.

Tugwell, Rexford G. (ed.) *The Trend of Economics*. Vols. I and II. Kennikat Press, Port Washington, NY, 1971 [1924].

Von Weizsacker, Carl F. "The Copenhagen Interpretation" in *Quantum Theory and Beyond.* Edited by T. Bastin. Cambridge University Press, Cambridge, England, 1971.

Von Mises, Ludwig. *Human Action: A treatise on Economics.* Fox and Wilkes, San Francisco, CA, [1949], 1996.

Weber, Max. "Science as a Vocation" in *From Max Weber*, edited by H.H. Gerth and C.W. Mills, New York, NY, 1946.

Weber, Max. *The Methodology of the Social Sciences*. Translated and edited by Edward A. Shils and Henry A. Finch. The Free Press, Glencoe, IL, 1949.

Weber, Max. *Critique of Stammler*. Translated by Guy Oakes. The Free Press, New York, NY, 1977.

Whorf, Benjamin Lee. *Language, Thought and Reality*. Edited by John B. Carroll. The M.I.T. Press, Cambridge, MA, 1956.

Wittgenstein, Ludwig. *Tractatus Logicus-Philosophicus*. Trans. by D.F. Pears and B.F. McGuinness. Macmillan, New York, NY 1961.

Wittgenstein, Ludwig. *Philosophical Investigations*. Trans. by G.E.M. Anscombe. Macmillan, New York, NY, 1953.

Wrong, Dennis H. *Population and Society*. Random House, NY, 1961.

Zeilinger, Anton. "Essential Quantum Entanglement" in *The New Physics for the Twenty-First Century*, edited by Gordon

Frazer, Cambridge University Press, Cambridge, UK, 2006. Pp. 257-67.

Zytkow, Jan M. and Willi Klosgen (Eds.) *Handbook of Data Mining and Knowledge Discovery.* Oxford University Press, NY, 2001.

Zytkow, Jan M. “Discovery of Regularities in Databases” in Proceedings *of the ML-92 Workshop on Machine Discovery (MD-92).*

Zytkow, Jan M. (Ed). National Institute for Aviation Research, Wichita State University KS, 1992.

Zytkow, Jan M. and Mohammed Quafafou (Eds.) *Principles of data Mining and Knowledge Discovery.* Springe-Verlagr, NY, 1998.